# Medium/Low Voltage Smart Grids

# Medium/Low Voltage Smart Grids

Editors

**Antonio Cataliotti**
**Dario Di Cara**
**Giovanni Artale**

MDPI • Basel • Beijing • Wuhan • Barcelona • Belgrade • Manchester • Tokyo • Cluj • Tianjin

*Editors*
Antonio Cataliotti
Università degli studi di Palermo
Italy

Dario Di Cara
National Research Council
of Italy
Italy

Giovanni Artale
Università degli studi di Palermo
Italy

*Editorial Office*
MDPI
St. Alban-Anlage 66
4052 Basel, Switzerland

This is a reprint of articles from the Special Issue published online in the open access journal *Energies* (ISSN 1996-1073) (available at: https://www.mdpi.com/journal/energies/special_issues/MLV_SG).

For citation purposes, cite each article independently as indicated on the article page online and as indicated below:

LastName, A.A.; LastName, B.B.; LastName, C.C. Article Title. *Journal Name* **Year**, *Volume Number*, Page Range.

**ISBN 978-3-0365-0470-4 (Hbk)**
**ISBN 978-3-0365-0471-1 (PDF)**

# Contents

# About the Editors

**Antonio Cataliotti** (Full Professor) received his M.S. and Ph.D. degrees in Electrical Engineering from University of Palermo, in 1992 and 1998, respectively. He is currently Full Professor in electrical and electronic measurements with the Department of Engineering of the University of Palermo. His current research interests include power quality measurements, power line communications, and smart grids.

**Dario Di Cara** (Ph.D.) received his M.S. and Ph.D. degrees in Electrical Engineering from University of Palermo, in 2005 and 2009, respectively. He is currently Researcher with the Institute of Marine Engineering of the Italian National Research Council, Palermo, Italy. His current research interests include power quality measurements, characterization of current transducers in non-sinusoidal conditions, power line communications, and smart grids.

**Giovanni Artale** (Ph.D.) received his M.S. degree in Electronic Engineering and Ph.D. degree in Electronic and Telecommunications Engineering from the University of Palermo, Palermo, Italy, in 2010 and 2014, respectively. He is currently Research Fellow with the Department of Engineering of the University of Palermo. His current research interests include low-frequency harmonic analysis algorithms, arc fault research methods, power line communications, and smart grids.

# Preface to "Medium/Low Voltage Smart Grids"

In the last decade, medium voltage (MV) and low voltage (LV) distribution networks are experiencing many changes due to the high increase in distributed generation from renewable energy sources, connection of new electric loads (such as electric vehicles), integration of energy storage systems, progressive participation of passive users to demand response strategies and introduction of new players in the energy market (including energy aggregators and virtual power plant, among others), and development of novel strategies for smart metering. In this panorama, distribution system operators need to revise their network management strategies, performing constant monitoring of the whole distribution network and interacting with distributed generators, energy storage systems, passive users, and energy aggregators.

This book is a collection of manuscripts proposing original and innovative solutions for accurate distributed monitoring systems and related innovative measurement instruments, distribution grid state forecast algorithms, active distribution networks with distributed generation, frequency and voltage control for network stability and quality of service, and communication systems to acquire distributed measurement data, send commands, and receive alarms. The introduction of these innovative solutions can pave the way for the effective transformation of MV and LV networks into smart grids.

The book aims to provide readers, Ph.D. students as well as research personnel and professional engineers with information not only on theoretical studies of the recent developments but also the practical application of the proposed solutions for smart grid applications both in LV and MV networks. The manuscripts of this book were developed by renowned researchers and specialists from around the world.

**Antonio Cataliotti, Dario Di Cara, Giovanni Artale**
*Editors*

*Article*

# Incremental Heuristic Approach for Meter Placement in Radial Distribution Systems

**Giovanni Artale [1], Antonio Cataliotti [1], Valentina Cosentino [1], Dario Di Cara [2,*], Salvatore Guaiana [1], Enrico Telaretti [1], Nicola Panzavecchia [2] and Giovanni Tinè [2]**

[1] Department of Engineering, Università degli Studi di Palermo, 90128 Palermo, Italy; giovanni.artale@unipa.it (G.A.); antonio.cataliotti@unipa.it (A.C.); valentina.cosentino@unipa.it (V.C.); salvatore.guaiana@unipa.it (S.G.); enrico.telaretti@unipa.it (E.T.)

[2] Institute of Marine Engineering (INM), National Research Council (CNR), 90146 Palermo, Italy; nicola.panzavecchia@cnr.it (N.P.); giovanni.tine@cnr.it (G.T.)

* Correspondence: dario.dicara@cnr.it

Received: 26 September 2019; Accepted: 15 October 2019; Published: 16 October 2019

**Abstract:** The evolution of modern power distribution systems into smart grids requires the development of dedicated state estimation (SE) algorithms for real-time identification of the overall system state variables. This paper proposes a strategy to evaluate the minimum number and best position of power injection meters in radial distribution systems for SE purposes. Measurement points are identified with the aim of reducing uncertainty in branch power flow estimations. An incremental heuristic meter placement (IHMP) approach is proposed to select the locations and total number of power measurements. The meter placement procedure was implemented for a backward/forward load flow algorithm proposed by the authors, which allows the evaluation of medium-voltage power flows starting from low-voltage load measurements. This allows the reduction of the overall costs of measurement equipment and setup. The IHMP method was tested in the real 25-bus medium-voltage (MV) radial distribution network of the Island of Ustica (Mediterranean Sea). The proposed method is useful both for finding the best measurement configuration in a new distribution network and also for implementing an incremental enhancement of an existing measurement configuration, reaching a good tradeoff between instrumentation costs and measurement uncertainty.

**Keywords:** optimal meter placement; smart grid; load flow analysis; Monte Carlo methods

## 1. Introduction

Modern power distribution grids are undergoing fundamental changes to their structure, thanks to the integration of distributed generators (DGs) and energy storages (DESs) from renewable energy sources, as well as the development of suitable communication systems and smart metering devices and infrastructures, all fostered by the political support of various countries and related standard requirements [1–7]. The transformation of traditional passive distribution networks into active grids (with bi-directional power flows and dynamic changes in grid operating conditions) requires the development of dedicated state estimation (SE) algorithms for real-time identification of the overall system state variables.

Traditionally, SE techniques make use of few actual measurements of medium-voltage (MV) branch voltages or current/power flows for collecting the input data of SE algorithms. The determination of the best possible combination of meters for distribution system monitoring is referred to as the optimal meter placement (MP) [8–14]. For the measurement of the SE input data, different kinds of measurement equipment can be used, i.e., phasor measurement units (PMUs), smart meters (SMs), power quality analyzers (PQAs) and so on, with different accuracy features and costs [15,16]. Missing data are integrated with pseudo-measurements, exploiting historical information or an a priori estimate of the relative power magnitude of each load [17–20]. Since pseudo-measurements are load estimates with high variance, the quality of the estimated state variables is dependent on the number of pseudo-measurements. If the SE errors are too high and do not ensure a reliable network control, additional real measurements are required. To this end, distribution system operators (DSOs) are required to choose the proper location, type and number of voltage or power measurements in the distribution network, in order to ensure suitable SE accuracy [21,22].

In theory, whenever an existing state estimator is upgraded or a new one is implemented, the best solution would be the redesign of the MP scheme for the whole network. However, in practical cases where an energy management system (EMS) and related measurement infrastructure are already implemented, this procedure is unfeasible (for economic reasons), and a reasonable solution is to implement an incremental enhancement of the existing measurement configuration. From this viewpoint, DSOs may be more interested in knowing how to improve their existing measurement infrastructure rather than in re-designing a new one. For this purpose, a meter placement method able to determine the proper locations of power measurements (PMs) can be very useful.

In this context, this paper proposes an IHMP method for placing PMs in radial distribution systems, for SE purposes. The proposed method focuses on branch power flows and uses a heuristic approach to select potential points for the location of bus PMs. The measurement points are identified with the aim of reducing uncertainty in branch power flow estimations, calculated using a Monte Carlo approach applied to a backward/forward (B/F) algorithm. Even though the technique does not explicitly optimize any objective function, the concept of an incremental approach is helpful in many practical situations. In particular, this technique is suitable in those cases where an EMS is already implemented and an enhancement of the measurement configuration is required.

The IHMP procedure was implemented and tested for a B/F load flow algorithm proposed by the authors, which allows the evaluation of MV power flows starting from low-voltage (LV) load measurements [23]. The integration of the proposed IHMP method with this load flow solution allows for an improvement in power flow estimation accuracy to be obtained, with limited costs for additional PMs. In fact, the proposed solution can be developed by using a low-cost measurement infrastructure that exploits PMs at LV level instead of those at MV level (as typically used in many literature MP and SE solutions). The IHMP method was tested in a real 25-bus MV radial distribution network. To validate the IHMP method's effectiveness, it is compared to that used in [24]; differently from [24], where the MP strategy was developed to minimize the uncertainty in power flow estimations at MV slack bus, the IHMP method aims at reducing the mean value of uncertainty in all MV branches by properly adding new PMs to existing ones. The obtained results show that the IHMP approach allows a significant reduction of uncertainties in most of the MV branches. Thus, the new contribution of this paper is the combined use of a recursive technique for meter placement based on uncertainty analysis and a new measurement approach for medium-voltage power flows starting from low-voltage load measurements. In this way, a good tradeoff between low cost and accuracy is reached, both when choosing a measurement configuration for a new distribution network and when implementing an incremental enhancement of an already existing measurement configuration. In fact, the LV measurement approach for MV power flow evaluation allows a reduction in cost because PMs are installed at the low-voltage side of the power transformer, while the IHMP technique allows suitable PM placement in order to obtain the target estimation accuracy. The proposed technique has the further positive feature of being based on a recursive procedure, using the Monte Carlo approach applied to the B/F load flow algorithm, which is itself also a recursive algorithm. With respect to more complicated algorithms based on genetic algorithms (GA) or particle swarm optimization (PSO), in fact, recursive algorithms have the great advantage of also being implementable in software with limited available libraries, such as those of Scada control centers.

The paper is organized as follows: Section 2 provides a state-of-the art overview of measurement systems and MP techniques for distribution grids. The IHMP method is described in Section 3, where a flow chart of the proposed algorithm is also provided. Section 4 presents the test results, which were obtained under the real test conditions of the MV radial distribution network of the Island of Ustica (where the measurement infrastructure for LV load power collection was implemented). Finally, Section 5 reports the conclusions of the work.

## 2. Meter Placement Techniques and Measurement Systems for Distribution Grids

Optimal meter placement is a topic of current research interest due to the large extension of distribution grids over the territory and the potential economic impact of smart grid applications and related needs for measurement infrastructures. Thus, several works can be found in the literature approaching the problem of determining the best MP configuration for a given power network. The problem has been addressed in literature, proposing different specific targets including system observability, installation/maintenance cost minimization, poor data detection capability, and SE accuracy [8–11]. To solve the issue, different optimization algorithms have been proposed, such as genetic algorithms (GA), particle swarm optimization (PSO) or heuristic techniques. As an example, in [12], a heuristic technique is proposed that identifies nodes in which to place a given number of voltage measurements, by reducing the standard deviation of voltages at nodes where measurements are not available. The main focus of [13], instead, is on reducing the uncertainties in specific critical points of the network, thus applying an incremental method to obtain the desired accuracy. In [14] the authors propose a two-stage PMU placement method: the first one is required to make the power system topologically observable, and the second one is proposed to check if the resulted placement leads to a full ranked measurement Jacobian. In [15], the attention is focused on current and power flow measurements and their impacts on branch current estimation accuracy. Further MP approaches are surveyed in [21,22].

As a general consideration, most of these studies are focused on the use of two types of measurements, i.e., bus voltage magnitude measurements and branch power flow measurements. The suggested measurement instruments are PMUs installed in MV nodes or power meters installed in MV branches. As regards this, it is known that the use of a high number of PMUs is not economically reasonable, even if such meters can allow the obtainment of higher accuracies. Thus, a common issue of most proposed methods is to achieve the goal of minimizing the total cost of the metering infrastructure on one hand and maximizing the estimation accuracy on the other. Typically, these two features conflict with each other: to obtain the lowest uncertainties at a measurement point, the best solution would be to use high-accuracy measurement instruments (typically PMUs) installed at MV level, thus coping with the costs of measurement instruments, MV transducers, and related installation costs. Thus, many distributed measurement system solutions proposed for MV power grids usually involve few MV metering points, whose information is integrated with that obtained by estimation algorithms [17–20]. This allows the reduction of total costs, but entails some drawbacks especially in terms of accuracy in the estimated quantities and also algorithm complexity.

In literature, some alternative solutions have been developed with the aim of reducing the total measurement infrastructure costs. For example, in [16], in order to find a good tradeoff between measurement equipment costs and SE accuracy, a hybrid solution is proposed based on the use of different kinds of meters. In [25], the use of LV smart meter measurements is also considered, in order to save the costs of additional MV measurement system installation. This approach is based on the assumption that real-time measurements of voltage and power are available in smart meters. On the other hand, this could be possible in a future generation of smart meters and related communication infrastructures, while already installed smart meters, used for billing purposes in many countries, typically provide only average measurements (usually over 15 min), which are collected with low updating frequency (usually once a day).

The authors also faced the issue of reducing the measurement infrastructure costs proposed in [23], a new measurement approach to perform load flow analysis in an MV network. The proposed distributed measurement system infrastructure is schematically represented in Figure 1.

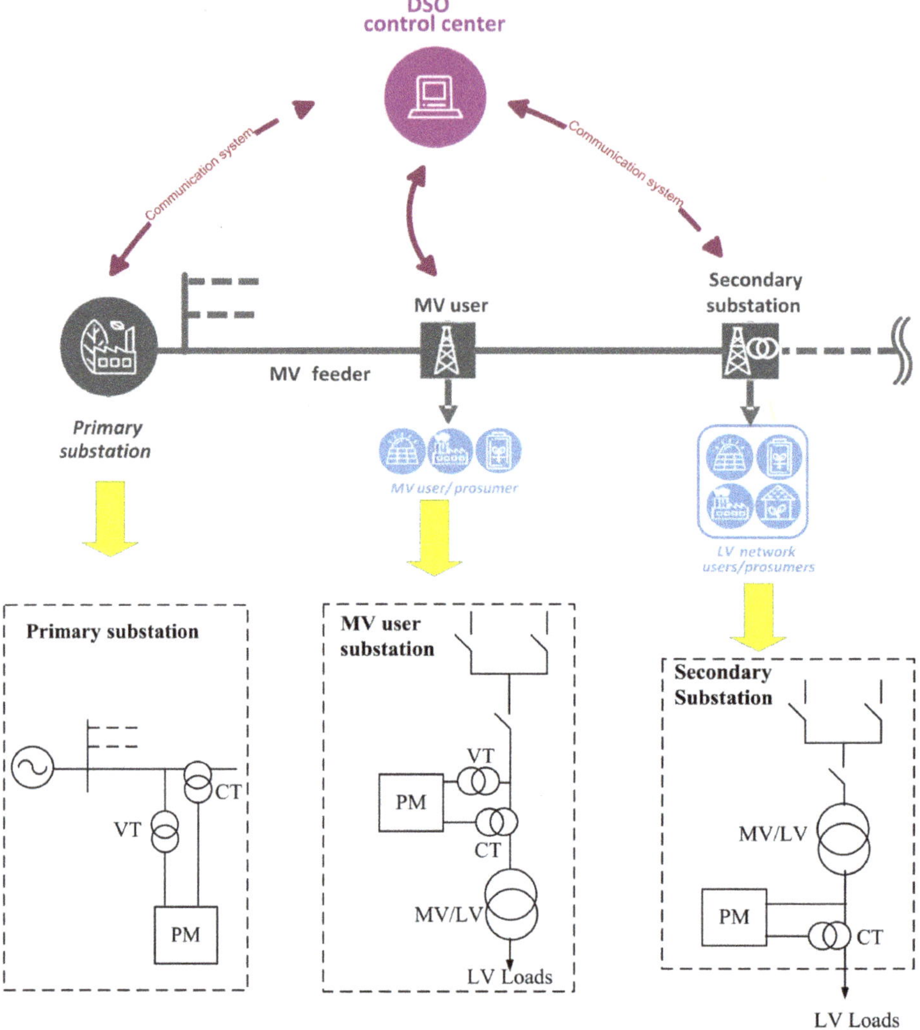


**Figure 1.** Measurement system architecture for medium-voltage (MV) network monitoring based on both MV and low-voltage (LV) load power measurements.

The proposed solution entails the use of PMs in both primary and secondary substations. In a primary substation, the PM measures the voltage of the slack bus and the total power drained by the MV feeder. Meanwhile, in the case of DSO and MV users, PMs measure the total load drained by the substation. In the case of primary and MV user substations, PMs are installed at MV level, using MV voltage and current transducers, (VT and CT, respectively). Meanwhile, in the case of a DSO's secondary substations, PMs are installed at LV level by using only LV current transducers. In this way, a reduction in installation costs is obtained because no MV transducers or related MV switchboard modifications are needed; moreover, current transducers for LV are less expensive than those for MV, and only a small number of customers would suffer an energy break, i.e., only those supplied by the secondary substation where the PM is installed, thus reducing the related DSO economic losses. Furthermore, PMs can also be already available in many secondary substations because they are usually installed by DSOs to discover eventual energy theft in LV networks. A backward/forward (B/F) algorithm is used to obtain the load flow in the MV network starting from these measurements. The algorithm takes into account the MV/LV power transformer losses, calculated starting from the measured load power and the power transformer rated data (i.e., short circuit and open circuit losses) [23]. Moreover, the algorithm also considers the line losses and voltage drops calculated starting from the rated data of MV lines. The load flow algorithm is implemented in the DSO control center. To collect the field measurement data, communication must be ensured between the DSO control center and the measurement points. In literature, different solutions are suggested to support smart grid applications, such as wireless, power line communications (PLC) or hybrid solutions, which can be also combined in an Internet of Things (IoT) platform [26–33]. On the other hand, currently, in MV networks DSOs deploy communication infrastructures only for automatic meter reading (AMR) and MV grid protection and supply restoration. The AMR server is usually installed in the DSO control center; it queries, via GSM (Global System for Mobile communications) or GPRS (General Packet Radio Service), the AMR concentrators of secondary substations. These AMR concentrators communicate via power line communications (PLC) to LV user smart meters. A supervisory system is also installed in the DSO control center for MV grid protection; it can control MV motorized switches, via GSM or GPRS, for fault isolation, protection or supply restoration purposes. In this framework, the idea of installing PMs at secondary substations perfectly fits with the described communication infrastructures, which can guarantee the real-time availability of measured data. In fact, PMs could be connected to the same communication system of AMR and MV networks and supply restoration systems, i.e., using installed GSM or GPRS systems. Moreover, this measurement infrastructure is also in line with the future implementation of faster communication links based on optic fibers, 5G or other wireless infrastructure. A similar solution is also adopted in the case study of the Ustica network, where a HiperLAN network was installed to connect the control center to the PMs of each secondary substation. More details on the infrastructure installed in the Ustica MV/LV network can be found in [4].

In light of the aforesaid considerations, the IHMP technique described in the next section, in conjunction with the abovementioned measurements and communication approach, can help to improve measurement uncertainty by suitably placing additional PMs in an existing measurement infrastructure, without entailing a high increase in the cost of measurement equipment. In fact, instead of using measurements at an MV level of voltage or current/power flow, the proposed measurement system suggests adding load power meters at the LV side of secondary substations, and the IHMP strategy allows the determination of where to locate these additional meters in order to guarantee the target uncertainty in estimated power flows all over the network.

## 3. Proposed Incremental Placement Technique

The purpose of the proposed IHMP technique is to determine the best location for PMs, exploiting a heuristic approach. Compared to the methods proposed in literature, the proposed approach is focused on branch power flows rather than nodal voltages. The measurement points are identified with the aim of reducing uncertainty in branch power flow estimations all over the network. The technique assumes that PMs are only located in certain nodes. In the remaining MV nodes, the load powers are estimated as pseudo-measurements. For sake of clarity, the method is firstly explained in a simple radial feeder, as shown in Figure 2. Meanwhile, its effectiveness in a real network application is shown in the next section.

The meter placement method starts by locating only one PM at the generic node $i$, whereas for the remaining MV nodes, pseudo-measurements are assumed. The uncertainties in all the branch power flows are obtained by applying a Monte Carlo procedure [34]: the load flow calculations are repeated 10,000 times, randomly changing the load powers at each node inside their Gaussian uncertainty distribution. In the nodes with PMs, the standard deviation of the distributions is that of the measurement instrument and the related transducers; meanwhile, in the nodes where pseudo-measurements are used, it is that of the estimation algorithm. At the end of the Monte Carlo procedure, the standard deviation of each branch power flow ($\sigma_j$) is estimated. By considering a cover factor of 3, which corresponds to a confidence level of 99.73%, the expanded uncertainty of the branch power flows can be obtained as:

$$U_j = 3\sigma_j \quad j = 1, 2 \ldots, N \tag{1}$$

where $N$ is the total number of branches of the feeder.

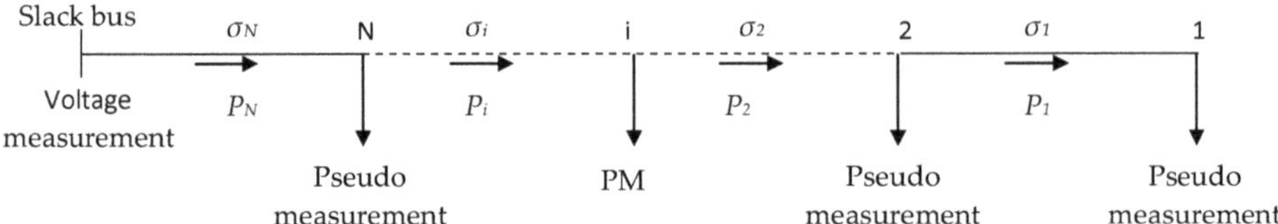


**Figure 2.** Simple radial feeder.

In order to assess to what extent branch power flow uncertainties vary when different locations of PMs are considered, the following index, $F_i$ (named the sensitivity index), is introduced:

$$F_i = \sqrt{\sum_{j=1}^{N} U_j^2} \tag{2}$$

where variable $i$ indicates the node where the meter is positioned, while variable j indicates the branch of the network. Starting from a given PM configuration (in terms of number and location of PMs), the sensitivity index $F_i$ gives an indication of the branch power flow uncertainties all over the network when an additional meter is located at the node $i$. Changing the node where the additional meter is placed, a different value of the index is obtained. Thus, the index allows the evaluation of the location of the meter that provides the lowest value of overall uncertainty.

A recursive procedure is then performed, by moving the meter in all the MV nodes, to identify the best position to locate the PM, i.e., the position that minimizes the sensitivity index, $F_i$. Once the first meter is placed, it is no longer moved. The algorithm continues looking for the position of a further meter. The sensitivity indexes are thus recalculated in the presence of the two meters: the first one is maintained fixed at the selected best location, and the second one is moved to the remaining nodes. At the end of the iteration procedure, the second meter is definitively located at the MV node that minimizes the sensitivity index. The procedure is thus repeated, locating other meters until an adequate accuracy level is achieved in all the branch power flows, i.e., when their uncertainties fall below a prefixed threshold. A flowchart of the proposed placement algorithm is shown in Figure 3. Firstly, a meter is added at node 1, whereas pseudo-measurements are assumed at the remaining MV nodes. The Monte Carlo procedure is applied to the load flow algorithm to evaluate the uncertainty of all branch power flows and thus to calculate the sensitivity index according to (2). This index is stored in memory. Then, the meter is removed from node 1 and relocated to node 2, and the Monte Carlo procedure is repeated to obtain the sensitivity index in this second case. This procedure is repeated iteratively, removing the meter from node $i$ and moving it to the following node $i$+1, until the end of the line is reached. The memorized sensitivity indexes are then compared, and the meter is definitively located in the node correspondent to the minimum sensitivity index. This procedure is again repeated, adding further meters and finding the best position in terms of the minimum sensitivity index at each iteration. The procedure is stopped when the power flow uncertainties are found to be below a chosen threshold in all the branches.

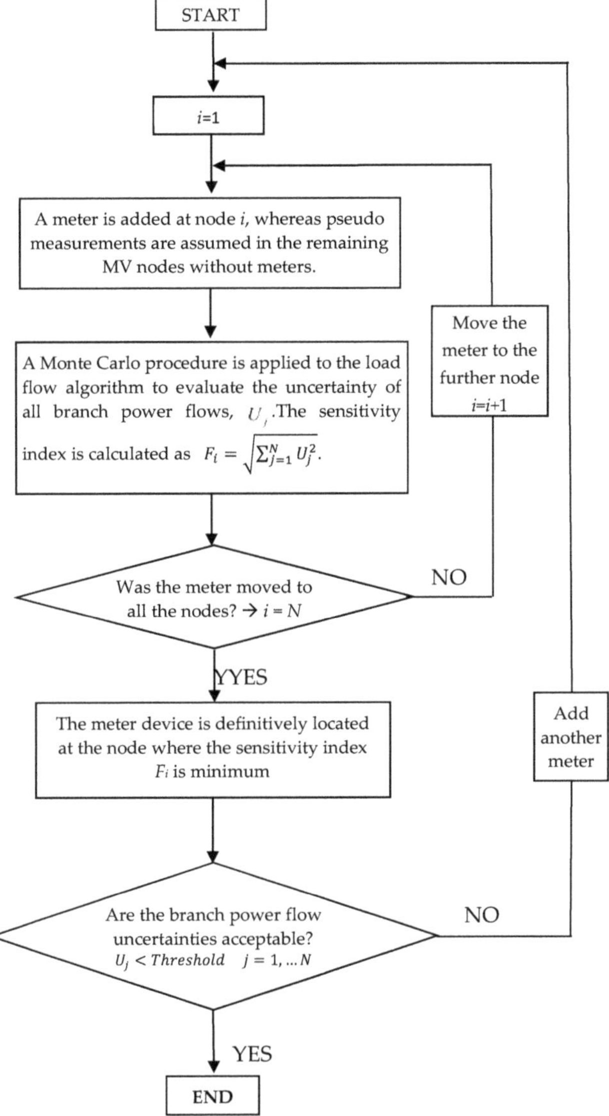


**Figure 3.** Flowchart of the proposed incremental placement algorithm.

## 4. Real Case Study Network and Simulation Results

The proposed meter placement algorithm was implemented and tested in the 25-bus MV radial distribution network of the Island of Ustica. The circuit diagram of the Ustica MV network is shown in Figure 4. The MV distribution network is composed of two radial feeders departing from a diesel power plant. The upper feeder, named Vittorio Emanuele feeder, supplies 13 secondary substations; the lower feeder, named AUSL feeder, supplies 11 substations. In the figure, for each MV branch (indicated with capital letters), the type (cable or overhead), the conductor core section and the total length are reported.

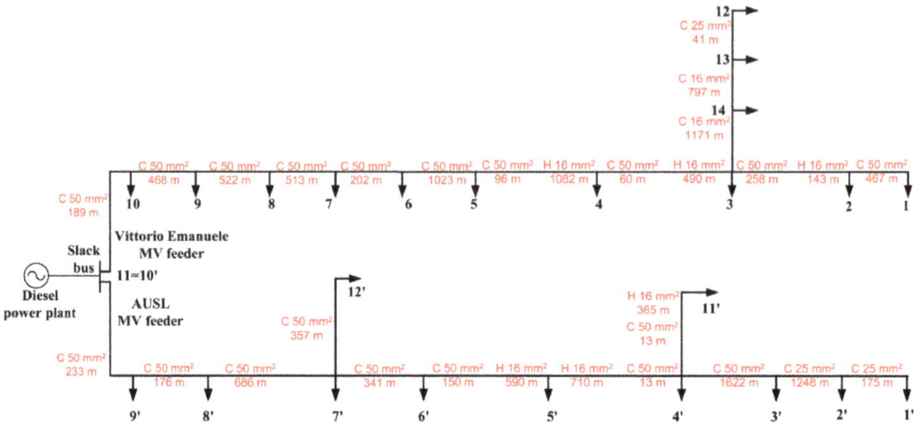

**Figure 4.** Circuit diagram of the Ustica MV network: the type (C = cable, H = overhead), the core section and the length are reported for each MV line.

The B/F load flow algorithm was implemented for the Ustica MV network. The MV bus-bars of the diesel power plant were chosen as slack bus for the load flow algorithm. The algorithm receives the following data as input variables:

- the load powers measured by the PMs;
- the load power estimations, derived from pseudo-measurements;
- the voltage at the slack bus;
- the line characteristics and the MV/LV power transformer parameters.

The B/F load flow algorithm is based on a recursive procedure: starting from the measured load powers, the algorithm iteratively updates branch power flows and node voltages using backward and forward sweeps on the network, respectively. The iterations are repeated until the convergence condition is found at slack bus, i.e., the difference between the calculated and measured voltages at the slack bus is below a certain threshold. This threshold was fixed according to the accuracy of the installed measurement system. More details on the B/F load flow algorithm can be found in [23].

In the bus where PMs are located, the load power measurements are affected by the uncertainty of the measurement system.

In MV users' secondary substations, the PM is usually installed at the MV side of power transformers, for billing purposes. In this case, the uncertainty of the PM is a composition of the PM and the related MV voltage and current transducer uncertainty, and thus it can be expressed as [18]:

$$u_{PM\_MV} = \frac{\sqrt{\eta_{CT}^2 + (100 \tan\theta \sin\epsilon_{CT})^2 + \eta_{VT}^2 + (100 \tan\theta \sin\epsilon_{VT})^2 + E_{P-PM}^2}}{\sqrt{3}} \tag{3}$$

where $\theta$ is the phase shift between voltage and current, $E_{P-PM}$ is the PM uncertainty in active power measurement, $\eta_{CT}$ and $\eta_{VT}$ are the ratio errors, and $\epsilon_{CT}$ and $\epsilon_{VT}$ are the phase displacements of the current and voltage transducers, respectively.

On the other hand, as explained in Section 2, in the DSO's secondary substation, the PM is installed at the LV side of power transformers. In this case, the PM uncertainty can be obtained from (3) without considering the terms related to VT errors.

To validate the proposed IHMP method, in the simulation study it was compared to that used in [24], showing how the new approach allows the reduction of uncertainties obtained in most of the branches. To perform the aforesaid comparison, the same reference load condition was used, which corresponds to the load power injection values shown in Table 1. This load condition is representative of the highest power flows measured in the summer period. In Table 1, the uncertainties are also reported for each node. They were calculated assuming 0.5 class instruments and transducers and also considering that CT errors are dependent on the current values and phase shift. In more detail, it is known that CT errors depend on the load currents. For example, for 0.5 class CTs, standard [35] fixes the maximum allowable ratio error $\eta_{CT\%} = \pm 0.5\%$ and maximum phase displacement $\varepsilon_{CT} = \pm 0.9$ *crad* at rated current. For current equal to 20% of the rated current, such error limits increase to $\eta_{CT\%} = \pm 0.75\%$ and $\varepsilon_{CT} = \pm 1.35$ *crad*. When currents are between 20% and 100% of the rated values, error limits are the linear interpolation between the two aforesaid limits. Thus, starting from the measured current and the rated value of the CT installed in each node, $\eta_{CT\%}$ and $\epsilon_{CT}$ are obtained and then PM uncertainties are calculated by using (3). More details on this procedure are reported in [36,37]. In the node without PM, a pseudo-measurement with an uncertainty of 100% is assumed in analogy of [24].

**Table 1.** Summer reference load condition.

| Vittorio Emanuele feeder | | | AUSL feeder | | |
|---|---|---|---|---|---|
| Node | P [kW] | $u_{PM}$ [kW] | Node | P [kW] | $u_{PM}$ [kW] |
| 1 | 8.964 | 0.15 | 1′ | 16.00 | 0.30 |
| 2 | 10.04 | 0.24 | 2′ | 116.0 | 1.33 |
| 3 | 38.52 | 0.44 | 3′ | 10.60 | 0.12 |
| 4 | 43.08 | 0.48 | 4′ | 29.28 | 0.34 |
| 5 | 56.00 | 0.61 | 5′ | 11.52 | 0.24 |
| 6 | 45.44 | 0.65 | 6′ | 21.76 | 0.51 |
| 7 | 249.6 | 2.94 | 7′ | 44.48 | 0.63 |
| 8 | 39.04 | 0.61 | 8′ | 124.5 | 1.40 |
| 9 [(a)] | 0 | | 9′ | 17.00 | 0.39 |
| 10 | 81.60 | 1.40 | 10′ [(b)] | - | |
| 11 [(b)] | - | | 11′ | 1.888 | 0.04 |
| 12 | 8.328 | 0.18 | 12′ | 55.68 | 0.81 |
| 13 | 51.00 | 0.42 | | | |
| 14 | 5.788 | 0.12 | | | |

[(a)] No low-voltage (LV) loads are connected to substation 9. [(b)] Nodes 11 and 10′ correspond to the slack bus.

Tables 2 and 3 show the values of the sensitivity indexes when the PM is moved between nodes, for the Vittorio Emanuele and the AUSL feeders, respectively. The sensitivity indexes refer to the total power of the feeder, which is 650.0 kW for the Vittorio Emanuele and 461.0 kW for the AUSL feeder. Table 2 shows that for the Vittorio Emanuele feeder, the sensitivity index has the lowest value when the first PM is located at node 7 (*Fi%* = 46.3%), the second PM at node 13 (*Fi%* = 40.5%), and so on (see values marked in bold). As regards the AUSL feeder, Table 3 shows that the first PM should be located at node 2′, where the sensitivity index is 54.4%, the second at node 8′, and so on (see values marked in bold).

**Table 2.** Sensitivity index as a percentage of the total power of the Vittorio Emanuele feeder when the power measurements (PMs) are moved between nodes.

| Number of PMs | Node | | | | | | | | | | | | |
|---|---|---|---|---|---|---|---|---|---|---|---|---|---|
| | 1 | 2 | 3 | 4 | 5 | 6 | 7 | 8 | 9 | 10 | 12 | 13 | 14 |
| 1 | 90.6 | 92.1 | 87.1 | 90.5 | 90.8 | 89.4 | **46.3** | 89.7 | 90.5 | 91.6 | 89.4 | 88.3 | 94.3 |
| 2 | 49.4 | 47.5 | 45.0 | 43.9 | 42.4 | 45.1 | DL [(a)] | 46.8 | 47.7 | 45.2 | 47.7 | **40.5** | 47.8 |
| 3 | 40.1 | 40.1 | 36.8 | 35.7 | **34.1** | 38.3 | DL | 38.7 | 40.3 | 38.7 | 40.2 | DL | 40.1 |
| 4 | 34.2 | 35.0 | 30.8 | **29.3** | DL | 32.0 | DL | 33.4 | 34.5 | 31.4 | 33.8 | DL | 34.3 |
| 5 | 29.4 | 29.4 | **23.7** | DL | DL | 24.6 | DL | 27.8 | 29.8 | 26.6 | 30.1 | DL | 29.7 |
| 6 | 24.8 | 24.2 | DL | DL | DL | **18.0** | DL | 21.7 | 24.0 | 20.4 | 24.7 | DL | 24.4 |
| 7 | 17.7 | 18.3 | DL | DL | DL | DL | DL | 15.1 | 18.3 | **13.6** | 17.2 | DL | 17.7 |
| 8 | 12.8 | 12.8 | DL | DL | DL | DL | DL | **8.2** | 13.5 | DL | 12.4 | DL | 13.3 |

[(a)] DL means that a PM is definitively located at the corresponding node, thus the sensitivity index is no longer calculated.

**Table 3.** Sensitivity index as a percentage of the total power of the AUSL feeder when the PMs are moved between nodes.

| Number of PMs | Node | | | | | | | | | | |
|---|---|---|---|---|---|---|---|---|---|---|---|
| | 1′ | 2′ | 3′ | 4′ | 5′ | 6′ | 7′ | 8′ | 9′ | 11′ | 12′ |
| 1 | 91.9 | **54.4** | 94.1 | 93.3 | 91.2 | 92.2 | 91.6 | 85.4 | 92.6 | 93.1 | 87.4 |
| 2 | 53.8 | DL [(a)] | 55.0 | 51.8 | 53.7 | 54.4 | 51.4 | **39.7** | 54.4 | 56.5 | 49.3 |
| 3 | 37.1 | DL | 38.1 | 34.7 | 38.3 | 36.8 | 34.1 | DL | 36.1 | 37.8 | **28.8** |
| 4 | 27.2 | DL | 28.5 | **24.0** | 28.0 | 27.8 | 24.4 | DL | 29.8 | 29.9 | DL |
| 5 | 22.3 | DL | 23.5 | DL | 24.3 | 22.5 | **17.1** | DL | 23.9 | 24.3 | DL |
| 6 | **13.4** | DL | 15.9 | DL | 16.2 | 14.3 | DL | DL | 17.0 | 18.1 | DL |
| 7 | DL | DL | 12.1 | DL | 12.5 | **9.7** | DL | DL | 12.9 | 13.6 | DL |
| 8 | DL | DL | **7.4** | DL | 7.7 | DL | DL | DL | 8.8 | 9.7 | DL |

[(a)] DL means that a PM is definitively located in the node, thus the sensitivity index is no longer calculated.

The uncertainties in the branch power flow estimations are shown in Figures 5 and 6 for the Vittorio Emanuele and AUSL feeders, respectively. The uncertainties are expressed as a percentage of the total power of the feeder. Each uncertainty curve corresponds to a given number and position of PMs. These PMs were located, one by one, in the MV nodes that ensure the minimum value of the sensitivity index. The graphs show that power flow uncertainties lower than 10% (desired accuracy level) can be obtained in all the branches by positioning PMs at only half of the nodes of each feeder.

Further simulations were performed to compare the branch power flow uncertainties obtained with the proposed method with those obtained by using the approach of [24]. The uncertainties obtained with the two methods are compared in Figures 7 and 8, in the case of seven PMs installed for each feeder. As can be seen, the power flux uncertainties obtained with the two methods are different. This is due to the different node selection for meter installations, which for the Vittorio Emanuele feeder is nodes 7, 13, 5, 4, 3, 6, and 10 for the proposed method and nodes 7, 10, 6, 8, 5, 4, and 3 for the second method, and for the AUSL feeder is nodes 2′, 8′, 12′, 4′, 7′, 1′, and 6′ for the proposed method and 8′, 2′, 12′, 7′, 6′, 9′, and 4′ for the second method. It should be noted that the two methods produce a different selection in both the chosen nodes and their order. With the proposed method, the medium value of uncertainty is lower than in the second method. On the other hand, the second method is optimized at the beginning of the feeder, thus a lower uncertainty is obtained in those branches. It should be noted that a very low difference is obtained at the beginning of the feeder with the second method, while a more significant reduction is obtained in the other branches with the proposed method.

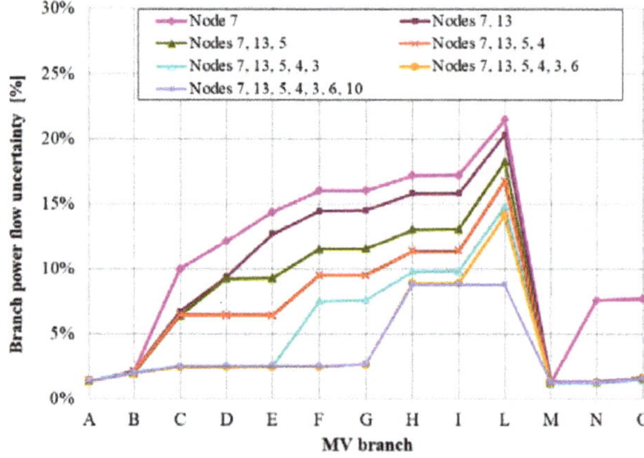


**Figure 5.** Branch power flow uncertainties (referring to the total power of the Vittorio Emanuele feeder) obtained when the number and position of PMs are changed.

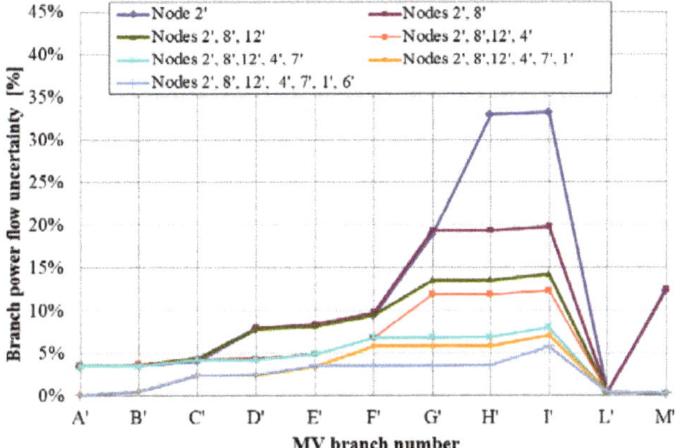


**Figure 6.** Branch power flow uncertainties (referring to the total power of the AUSL feeder) obtained when the number and position of PMs are changed.

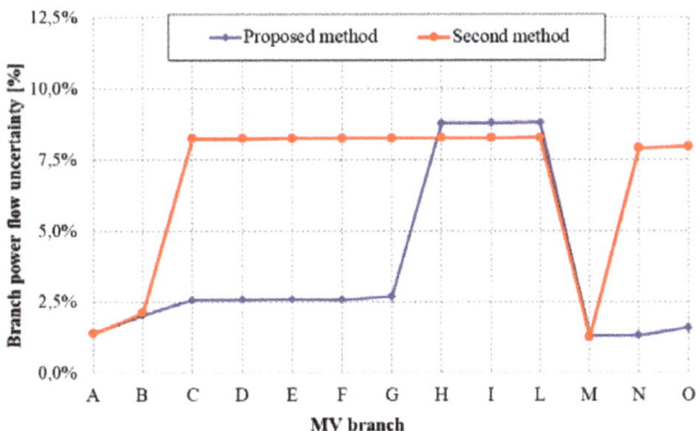


**Figure 7.** Branch power flow uncertainty with seven PMs located in the Vittorio Emanuele feeder: comparison with a second placement method.

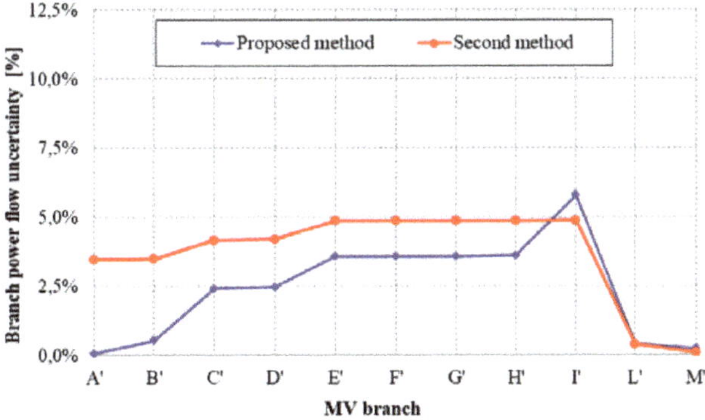


**Figure 8.** Branch power flow uncertainty with seven PMs located in the AUSL feeder: comparison with a second placement method.

Using the same PM configuration, the branch power flow uncertainties were evaluated in a further load condition, registered on 1 November at 12:30 (see Table 4). This load condition was chosen since it is representative of the lowest power flows observed in the Ustica MV network (in the winter period). The obtained uncertainties are shown in Figures 9 and 10. They are compared to those obtained in the summer load condition. As can be seen, the power flow uncertainty is lower than the desired accuracy level in all the branches, meaning that the proposed meter placement procedure ensures an adequate power flow uncertainty level for different load conditions.

**Table 4.** Winter reference load condition.

| Node | Vittorio Emanuele Feeder P [kW] | Node | AUSL Feeder P [kW] |
|---|---|---|---|
| 1 | 0.80 | 1′ | 5.00 |
| 2 | 1.10 | 2′ [(a)] | 0.00 |
| 3 | 17.80 | 3′ | 2.90 |
| 4 | 17.30 | 4′ | 7.60 |
| 5 | 8.40 | 5′ | 3.90 |
| 6 | 16.20 | 6′ | 5.20 |
| 7 | 52.10 | 7′ | 7.10 |
| 8 | 5.60 | 8′ | 48.50 |
| 9 [(a)] | 0.00 | 9′ | 10.50 |
| 10 | 56.80 | 10′ [(b)] | 0 |
| 11 [(b)] | 0 | 11′ [(a)] | 0.00 |
| 12 | 5.20 | 12′ | 45.50 |
| 13 | 42.00 | | |
| 14 | 3.10 | | |

[(a)] No LV loads are connected to substations 9, 2′ and 11′ in winter. [(b)] Nodes 11 and 10′ correspond to the slack bus.

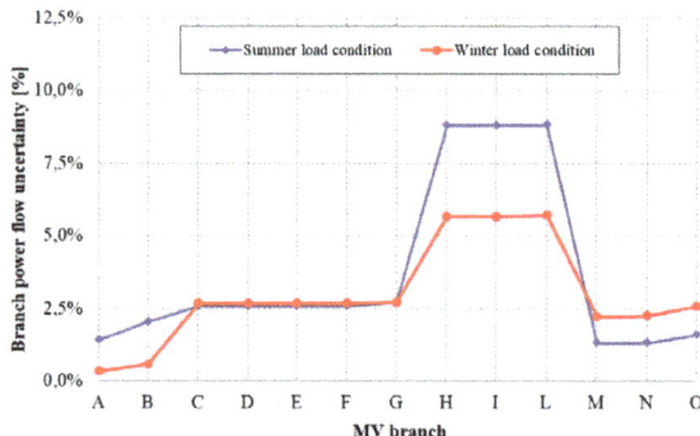


**Figure 9.** Branch power flow uncertainty with seven PMs located in the Vittorio Emanuele feeder: comparison between summer and winter load conditions.

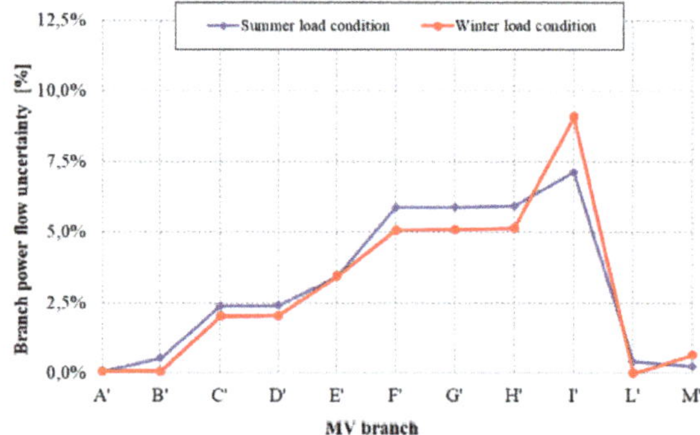


**Figure 10.** Branch power flow uncertainty with six PMs located in the AUSL feeder: comparison between summer and winter load conditions.

## 5. Discussion

The proposed method, minimizing the sensitivity index, allows there to be a minimal mean value of uncertainty all over the network. To verify that the obtained solutions are optimal, some further tests were performed to investigate if, starting from a given configuration (which ensures an uncertainty value lower than the desired threshold), one of the placed meters could be removed and still have the condition satisfied. For example in the following, the results related to one of the test cases of the Vittorio Emanuele feeder are presented, i.e., the metering condition of the seventh iteration, which allowed the satisfaction of the 10% threshold in all the branches. Starting from this metering condition, a meter was removed from one of the seven nodes and the Monte Carlo procedure was performed again to calculate the power flow uncertainties and the sensitivity index. This was done for all seven nodes. The branch power flow uncertainties obtained in these tests are shown in Figure 11. As can be seen, none of the selected cases with six meters satisfy the 10% condition, thus confirming the need for a seventh meter as requested by the proposed method. Moreover, the indexes found for each removed node are reported in Table 5. As can be seen, none of the tested cases allows the obtainment of a sensitivity index lower than that found with the proposed solution at the sixth iteration, thus confirming the optimality of the solution.

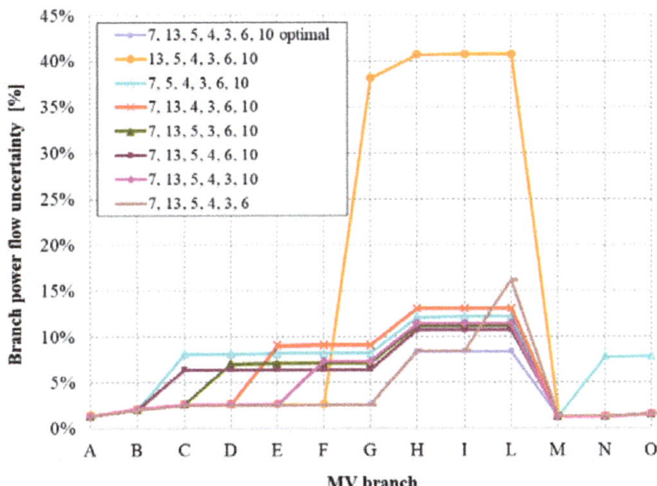


**Figure 11.** Branch power flow uncertainties obtained when one PM is removed from the seven-meter optimal condition found for the Vittorio Emanuele feeder.

**Table 5.** Sensitivity index when one PM is removed from the seven-meter optimal condition found for the Vittorio Emanuele feeder.

| Nodes with Meters | Node Removed | $F_i$ Index |
|---|---|---|
| 7, 13, 5, 4, 3, 6, 10 Optimal seventh iteration | None | 13.6% |
| 13, 5, 4, 3, 6, 10 | 7 | 68.4% |
| 7, 5, 4, 3, 6, 10 | 13 | 25.7% |
| 7, 13, 4, 3, 6, 10 | 5 | 23.8% |
| 7, 13, 5, 3, 6, 10 | 4 | 20.7% |
| 7, 13, 5, 4, 6, 10 | 3 | 20.2% |
| 7, 13, 5, 4, 3, 10 | 6 | 19.6% |
| 7, 13, 5, 4, 3, 6 Optimal seventh iteration | 10 | 18.0% |

A further test was performed to verify the optimality of the proposed solution. Performing an extra iteration of the method (the eighth), a further meter would be added to node 8, as demonstrated by Table 2. Thus, the further tests consisted in adding the meter to the eighth node and removing one meter from the nodes by the seventh iteration. The branch power flow uncertainties obtained in these tests are shown in Figure 12. As can be seen, some of the selected cases with seven meters satisfy the 10% threshold but none of them has a lower mean value of uncertainty in the whole network than that found for the optimal seven-meter solution. This is demonstrated by the $F_i$ index found for each case, reported in Table 6. As can be seen, none of the cases allows the obtainment of a lower value of the Fi index with respect to that found for the seventh iteration, thus confirming the optimality of the solution.

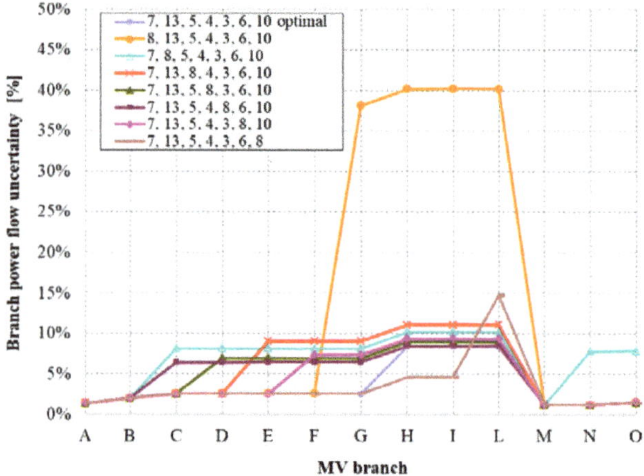


**Figure 12.** Branch power flow uncertainties obtained when one PM is added to node 8 and one PM is removed from the seven-meter optimal condition found for the Vittorio Emanuele feeder.

**Table 6.** Sensitivity index when one PM is added to node 8 and one PM is removed from the seven-meter optimal condition found for the Vittorio Emanuele feeder.

| Nodes with Meters | Node Substituted with Node 8 | $F_i$ Index |
|---|---|---|
| 7, 13, 5, 4, 3, 6, 10 Optimal seventh iteration | None | 13.6% |
| 8, 13, 5, 4, 3, 6, 10 | 7 | 67.8% |
| 7, 8, 5, 4, 3, 6, 10 | 13 | 23.8% |
| 7, 13, 8, 4, 3, 6, 10 | 5 | 21.5% |
| 7, 13, 5, 8, 3, 6, 10 | 4 | 18.1% |
| 7, 13, 5, 4, 8, 6, 10 | 3 | 17.8% |
| 7, 13, 5, 4, 3, 8, 10 | 6 | 17.1% |
| 7, 13, 5, 4, 3, 6, 8 | 10 | 15.1% |

## 6. Conclusions

This paper proposes a new IHMP method to locate PMs for SE purposes in radial distribution systems. In more detail, the proposed approach is focused on branch power flow estimation, and it allows the determination of the minimum number and best position of load power meters in order to achieve a target accuracy in the branch power flow estimation. The proposed measurement technique assumes that power meters are located in a limited number of nodes, whereas in the remaining grid nodes the load powers are estimated as pseudo-measurements, with a given uncertainty. A sensitivity index is defined, which takes into account the branch power flow uncertainty all over the network; this ensures that, at any branch, the power flow is evaluated with an uncertainty value below a prefixed threshold.

The IHMP procedure was implemented for a backward/forward load flow algorithm proposed by the authors, which allows the evaluation of MV branch power flows starting from LV load power measurements. The integration of the proposed IHMP method with this load flow solution allows the obtainment of an improvement in power flow estimation accuracy, with limited costs for PMs. In fact, the proposed solution can be developed by using a low-cost measurement infrastructure that exploits PMs at LV level instead of those at MV level. The proposed method was implemented and tested in real case of the MV radial distribution network of the Island of Ustica, under different load conditions. The obtained results confirm that the method is useful to find the measurement configuration that guarantees a low uncertainty all over the network. In the case under study, the proposed meter placement procedure ensures an adequate power flow uncertainty level when PMs are installed in about half of the grid nodes.

The proposed IHMP technique is incremental, thus it allows not only for the best measurement configuration to be found in a new distribution network, but also allows the determination of how to obtain an enhancement of an existing measurement configuration. Both applications can be of practical interest for DSOs, as they may be interested in either installing a new measurement infrastructure or enhancing an already owned one. In both cases, the possibility of using measurements at LV level can increase the cost-effectiveness of the measurement infrastructure. In fact, thanks to the use of LV measurements instead of MV ones and the use of the proposed IHMP method, a good tradeoff can be reached between low cost and accuracy: the cost is reduced because of the choice to install PMs at the low-voltage side of the power transformer, while the IHMP technique allows suitable PM placement in order to obtain the target estimation accuracy.

**Author Contributions:** Conceptualization, A.C., V.C., D.D.C. and G.T.; Data curation, G.A., S.G., N.P. and E.T.; Investigation, A.C., V.C., D.D.C. and G.T.; Methodology, G.A., A.C., V.C., D.D.C., S.G., N.P., E.T. and G.T.; Software, G.A., S.G., N.P. and E.T.; Supervision, A.C., V.C., D.D.C. and G.T.; Validation, G.A., A.C., V.C., D.D.C., S.G., N.P., E.T. and G.T.; Writing—original draft, G.A., S.G., N.P. and E.T.; Writing—review & editing, A.C., V.C., D.D.C. and G.T.

**Funding:** This research received no external funding

**Acknowledgments:** The authors wish to thank the local DSO of Ustica (Impresa Elettrica D'Anna e Bonaccorsi S.R.L.) for their support and the collected network parameters and measurement data.

**Conflicts of Interest:** The authors declare no conflict of interest.

## References

1. De Villena, M.M.; Fonteneau, R.; Gautier, A.; Ernst, D. Evaluating the evolution of distribution networks under different regulatory frameworks with multi-agent modelling. *Energies* **2019**, *12*, 7. [CrossRef]
2. TT Tran, Q.; Luisa Di Silvestre, M.; Riva Sanseverino, E.; Zizzo, G.; Pham, T.N. Driven Primary Regulation for Minimum Power Losses Operation in Islanded Microgrids. *Energies* **2018**, *11*, 2890. [CrossRef]
3. Osório, G.J.; Shafie-khah, M.; Coimbra, P.D.L.; Lotfi, M.; Catalão, J.P.S. Distribution System Operation with Electric Vehicle Charging Schedules and Renewable Energy Resources. *Energies* **2018**, *11*, 3117. [CrossRef]
4. Artale, G.; Cataliotti, A.; Cosentino, V.; Di Cara, D.; Guaiana, S.; Panzavecchia, N.; Tinè, G. Real Time Power Flow Monitoring and Control System for Microgrids Integration in Islanded Scenarios. *IEEE Trans. Ind. Appl.* **2019**. [CrossRef]
5. Christoforidis, C.G.; Panapakidis, P.I.; Papadopoulos, A.T.; Papagiannis, K.G.; Koumparou, I.; Hadjipanayi, M.; Georghiou, E.G. A Model for the Assessment of Different Net-Metering Policies. *Energies* **2016**, *9*, 262. [CrossRef]
6. Crotti, G.; Delle Femine, A.; Gallo, D.; Giordano, D.; Landi, C.; Luiso, M. Measurement of the Absolute Phase Error of Digitizers. *IEEE Trans. Instrum. Meas.* **2019**, *68*, 1724–1731. [CrossRef]
7. Dusonchet, L.; Favuzza, S.; Massaro, F.; Telaretti, E.; Zizzo, G. Technological and legislative status point of stationary energy storages in the EU. *Renew. Sustain. Energy Rev.* **2019**, *101*, 158–167. [CrossRef]

8. Chen, X.; Lin, J.; Wan, C.; Song, Y.; You, S.; Zong, Y.; Guo, W.; Li, Y. Optimal Meter Placement for Distribution Network State Estimation: A Circuit Representation Based MILP Approach. *IEEE Trans. Power Syst.* **2016**, *31*, 4357–4370. [CrossRef]
9. Xygkis, T.C.; Korres, G.N. Optimized measurement allocation for power distribution systems using mixed integer sdp. *IEEE Trans. Instrum. Meas.* **2017**, *66*, 2967–2976. [CrossRef]
10. Damavandi, M.G.; Krishnamurthy, V.; Martí, J.R. Robust Meter Placement for State Estimation in Active Distribution Systems. *IEEE Trans. Smart Grid* **2015**, *6*, 1972–1982. [CrossRef]
11. Yao, Y.; Liu, X.; Li, Z. Robust Measurement Placement for Distribution System State Estimation. *IEEE Trans. Sustain. Energy* **2019**, *10*, 364–374. [CrossRef]
12. Shafiu, A.; Jenkins, N.; Strbac, G. Measurement location for state estimation of distribution networks with generation. *IEE Proc. Gener. Transm. Distrib.* **2005**, *152*, 240–246. [CrossRef]
13. Xiang, Y.; Ribeiro, P.F.; Cobben, J.F.G. Optimization of state estimator-based operation framework including measurement placement for medium voltage distribution grid. *IEEE Trans. Smart Grid* **2014**, *5*, 2929–2937. [CrossRef]
14. Sodhi, R.; Srivastava, S.C.; Singh, S.N. Optimal PMU placement method for complete topological and numerical observability of power system. *Electr. Power Syst. Res.* **2010**, *80*, 1154–1159. [CrossRef]
15. Pau, M.; Pegoraro, P.A.; Monti, A.; Muscas, C.; Ponci, F.; Sulis, S. Impact of Current and Power Measurements on Distribution System State Estimation Uncertainty. *IEEE Trans. Instrum. Meas.* **2019**. [CrossRef]
16. Prasad, S.; Kumar, D.M.V. Trade-offs in PMU and IED deployment for active distribution state estimation using multi-objective evolutionary algorithm. *IEEE Trans. Instrum. Meas.* **2018**, *67*, 1298–1307. [CrossRef]
17. Pokhrel, B.R.; Bak-Jensen, B.; R Pillai, J. Integrated Approach for Network Observability and State Estimation in Active Distribution Grid. *Energies* **2019**, *12*, 2230. [CrossRef]
18. Cataliotti, A.; Cervellera, C.; Cosentino, V.; Di Cara, D.; Gaggero, M.; Maccio, D.; Marsala, G.; Ragusa, A.; Tine, G. An Improved Load Flow Method for MV Networks Based on LV Load Measurements and Estimations. *IEEE Trans. Instrum. Meas.* **2019**, *68*, 430–438. [CrossRef]
19. Soares, T.M.; Bezerra, U.H.; Tostes, M.E.L. Full-Observable Three-Phase State Estimation Algorithm Applied to Electric Distribution Grids. *Energies* **2019**, *12*, 1327. [CrossRef]
20. Brinkmann, B.; Member, S.; Negnevitsky, M.; Member, S. A Probabilistic Approach to Observability of Distribution Networks. *IEEE Trans. Power Syst.* **2017**, *32*, 1169–1178. [CrossRef]
21. Dehghanpour, K.; Wang, Z.; Wang, J.; Yuan, Y.; Bu, F. A survey on state estimation techniques and challenges in smart distribution systems. *IEEE Trans. Smart Grid* **2019**, *10*, 2312–2322. [CrossRef]
22. Lekshmana, R.; Padmanaban, S.; Mahajan, S.B.; Ramachandaramurthy, V.K.; Holm-Nielsen, J.B. Meter placement in power system network—A comprehensive review, analysis and methodology. *Electronics* **2018**, *7*, 329. [CrossRef]
23. Cataliotti, A.; Cosentino, V.; Di Cara, D.; Russotto, P.; Telaretti, E.; Tinè, G. An Innovative Measurement Approach for Load Flow Analysis in MV Smart Grids. *IEEE Trans. Smart Grid* **2016**, *7*, 889–896. [CrossRef]
24. Cataliotti, A.; Cosentino, V.; Di Cara, D.; Tinè, G. LV Measurement Device Placement for Load Flow Analysis in MV Smart Grids. *IEEE Trans. Instrum. Meas.* **2016**, *65*, 999–1006. [CrossRef]
25. Pau, M.; Patti, E.; Barbierato, L.; Estebsari, A.; Pons, E.; Ponci, F.; Monti, A. Design and Accuracy Analysis of Multilevel State Estimation Based on Smart Metering Infrastructure. *IEEE Trans. Instrum. Meas.* **2019**, 1–13. [CrossRef]
26. Al-Turjman, F.; Abujubbeh, M. IoT-enabled smart grid via SM: An overview. *Future Gener. Comput. Syst.* **2019**, *96*, 579–590. [CrossRef]
27. Sanduleac, M.; Lipari, G.; Monti, A.; Voulkidis, A.; Zanetto, G.; Corsi, A.; Toma, L.; Fiorentino, G.; Federenciuc, D. Next Generation Real-Time Smart Meters for ICT Based Assessment of Grid Data Inconsistencies. *Energies* **2017**, *10*, 857. [CrossRef]
28. Marcon, P.; Szabo, Z.; Vesely, I.; Zezulka, F.; Sajdl, O.; Roubal, Z.; Dohnal, P. A Real Model of a Micro-Grid to Improve Network Stability. *Appl. Sci.* **2017**, *7*, 757. [CrossRef]
29. Rinaldi, S.; Pasetti, M.; Sisinni, E.; Bonafini, F.; Ferrari, P.; Rizzi, M.; Flammini, A. On the Mobile Communication Requirements for the Demand-Side Management of Electric Vehicles. *Energies* **2018**, *11*, 1220. [CrossRef]

30. Artale, G.; Cataliotti, A.; Cosentino, V.; Di Cara, D.; Fiorelli, R.; Guaiana, S.; Panzavecchia, N.; Tinè, G. A new low cost power line communication solution for smart grid monitoring and management. *IEEE Instrum. Meas. Mag.* **2018**, *21*, 29–33. [CrossRef]
31. Ouissi, S.; Ben Rhouma, O.; Rebai, C. Statistical modeling of mains zero crossing variation in powerline communication. *Meas. J. Int. Meas. Confed.* **2016**, *90*, 158–167.
32. Kabalci, Y.; Kabalci, E. Modeling and analysis of a smart grid monitoring system for renewable energy sources. *Sol. Energy* **2017**, *153*, 262–275. [CrossRef]
33. Rinaldi, S.; Bonafini, F.; Ferrari, P.; Flammini, A.; Sisinni, E.; Di Cara, D.; Panzavecchia, N.; Tine, G.; Cataliotti, A.; Cosentino, V.; et al. Characterization of IP-Based communication for smart grid using software-defined networking. *IEEE Trans. Instrum. Meas.* **2018**, *67*, 2410–2419. [CrossRef]
34. Joint Committee for Guides in Metrology (JCGM). *Evaluation of Measurement Data—Supplement 1 to the 'Guide to the Expression of Uncertainty in Measurement'—Propagation of Distributions Using a Monte Carlo Method, 101:2008, Ed.1*; JCGM, BIPM: Sevres, France, 2008.
35. IEC Standard. *Instrument Transformers—Part 2: Additional Requirements for Current Transformers IEC Standard 61869-2*; IEC Standard: Geneva, Switzerland, 2012.
36. Cataliotti, A.; Cosentino, V.; Di Cara, D.; Guaiana, S.; Nuccio, S.; Panzavecchia, N.; Tinè, G. Measurement uncertainty impact on simplified load flow analysis in MV smart grids. In Proceedings of the 2018 IEEE International Instrumentation and Measurement Technology Conference (I2MTC), Houston, TX, USA, 14–17 May 2018; pp. 1354–1359.
37. Cataliotti, A.; Cosentino, V.; Di Cara, D.; Telaretti, E.; Tinè, G. Uncertainty evaluation of a backward/forward load flow algorithm for a MV smart grid. In Proceedings of the IEEE International Instrumentation Measurement Technology Conference, Pisa, Italy, 11–14 May 2015; pp. 1279–1284.

Article

# Evaluation of Different Development Possibilities of Distribution Grid State Forecasts

**Jessica Hermanns *, Marcel Modemann, Kamil Korotkiewicz, Frederik Paulat, Kevin Kotthaus, Sven Pack and Markus Zdrallek**

Institute of Power System Engineering, University of Wuppertal, 42119 Wuppertal, Germany; marcel.modemann@uni-wuppertal.de (M.M.); kamil.korotkiewicz@uni-wuppertal.de (K.K.); frederik.paulat@uni-wuppertal.de (F.P.); kevin.kotthaus@uni-wuppertal.de (K.K.); sven.pack@uni-wuppertal.de (S.P.); zdrallek@uni-wuppertal.de (M.Z.)

* Correspondence: jessica.hermanns@uni-wuppertal.de

Received: 7 March 2020; Accepted: 9 April 2020; Published: 13 April 2020

**Abstract:** The number of renewable energy systems is still increasing. To reduce the worldwide $CO_2$ emissions, there will be even more challenges in the distribution grids like currently upcoming charging stations or heat pumps. All these new electric systems in the low voltage (LV) and medium voltage (MV) levels are characterized by an unsteady behavior. To monitor and predict the behavior of these new flexible systems, a grid state forecast is needed. This software tool calculates wind, photovoltaic, and load forecasts. These power forecasts are already in the focus of research, but there are some specific use cases, which require a more specific solution. To get a variously applicable software tool, different new functions to improve an already existing grid state forecast tool were developed and evaluated. For example, it will be proofed if a grid state forecast tool can be improved by calculating the number or the base load of the loads in grid areas by just one available measurement. Another big subject exists in the exchange of forecast information between different voltage levels. How this can be realized and how big the effect on the forecast quality is, will be analyzed. The results of these evaluations will be shown in this paper.

**Keywords:** grid state forecast; smart grids; distribution grids; congestion management

---

## 1. Introduction

To achieve the aim of energy supply without pollutant emissions, what could be reached until 2050 according to [1], a further expansion of renewable energy systems is needed. Due to that, there will be more and more non time-discrete loads and generators in the distribution grids, which stress the grids in various ways. In a few times, there could occur different improbable load and generation situations at the same time, which could cause grid congestions. One example of these unfavorable overlaps could be a situation of less load and much generation. To prevent these rare situations, a grid state forecast is favorable as information source for the distribution system operator (DSO).

Thereby the grid congestion is predicted soon enough to react by a local flexibility market presented in [2,3] or by a preventive controlling smart grid system. In these concepts, the predicted voltage and current values of all nodes and branches are evaluated for a voltage boundary violation or a cable overload. Then, the according algorithm calculates how the grid congestion could be prevented. This could be done by load shifting or throttling down a generation system for example. That these preventive concepts of using flexibility can be more effective and less expensive than conventional grid enhancement is shown [4].

In the literature, there are some approaches for grid state forecasts shown in [5–7], but none of them seem usable for this specific use case. The biggest common disadvantage is the dependence of available measurements. For a local flexibility market or a preventive controlling smart grid system, a high temporal resolution of the forecasts is mandatory, so that the runtime is not in real-time, but in the minute range. The forecast has to be applicable for small loads like households or greater aggregated consumptions like medium voltage (MV)/low voltage (LV) transformers. The forecast should be capable to process topology changes and changes of the supply task. The grid state forecast should also be robust against communication failures and restarts, which means that the forecast should calculate values also without actual or previous measurements by means of a replacement value creation.

The bottom-up approach, which will be presented in this paper, can handle different situations, which can occur in a real-time use of a grid state forecast. The bottom-up approach is constructed by different levels of available information. The already existing grid state forecasts from the literature, like they are presented in [5–7], are all constructed to handle many measurements from the grid and there is no solution to handle grids without measurements. Many load forecasts, like [8–10], need every load node to be measured to predict it. This is not realistic and will not happen next time in the low and medium voltage grids. The developed bottom-up approach can handle situations with no measurements or with just one time step measured, for example, after a restart of the system. The load forecast differs if there are schedules, smart meter data, load types, previous voltage drops, or current measurements. The bottom-up approach allows it to exchange forecasts between different voltage levels, which is important to improve the forecast quality.

To predict voltage and current values of high quality, the grid state forecast has to take into account every load and generation in the grid, so that a modular bottom-up approach is used, which is shown in Figure 1.

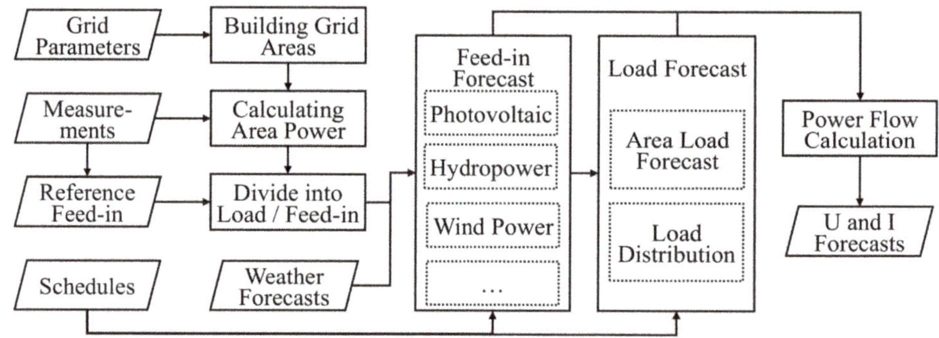


**Figure 1.** Procedure of a grid state forecast by a bottom-up approach.

Assuming the bottom-up approach, the grid is separated into different grid areas. These are defined by the positions of the sensors in the grid. One example is shown in Figure 2. By the difference between incoming and outgoing currents out of the grid area, the obtained currents in the grid area can be calculated. The calculated grid area power can then be separated into load and generation by some reference generation systems, which are measured. By this procedure, for example, the generation of all photovoltaic systems in the grid can be calculated by one measured photovoltaic system. This method is inspired by the estimation of feed-in in smart grid systems as in [11].

The generation systems are all predicted on their own by several models for each type. For photovoltaic systems, the characteristics of the module, the geographical position, and the date are needed to calculate the theoretically possible feed-in of the considered day. This photovoltaic power can then be calculated with external weather forecasts. Other feeder models, as for example the wind forecast, are similarly structured. For storage systems, biogas systems, and other generation systems with plannable generation, the availability of schedules is assumed to be integrated in a modular manner into the forecast concept.

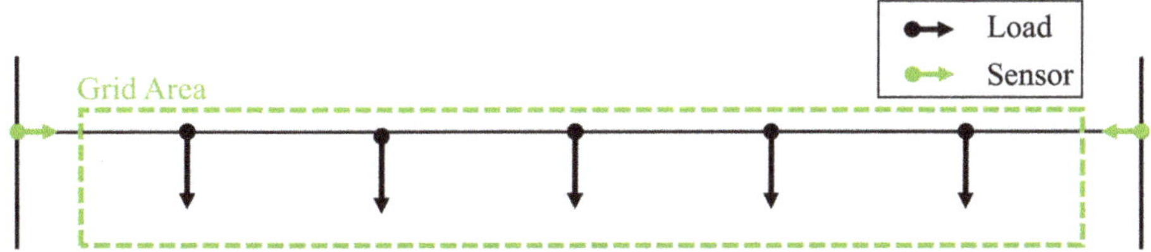


**Figure 2.** Example of a grid area defined by two sensors.

These forecasts can be improved by some single measurements at the reference systems. How the measurements can improve the forecasts is shown in [12]. How it is realized in the presented grid state forecast shows Equation (1). The factor for considering the measurements $\underline{f}_T$ is the result of the difference between the real apparent power $\underline{\mathbf{s}} := (\underline{s}_1, \ldots, \underline{s}_n)^T$ and the predicted apparent power $\underline{\mathbf{s}}^p := (\underline{s}_1^p, \ldots, \underline{s}_n^p)^T$ of $n$ time steps. The factor of the previous forecast $\underline{f}_{T-1}$ is included with the weighting factor $w$, which can be values between zero and one. Which value for the weighting factor gets the smallest forecast error has to be analyzed with more data. By this approach, the forecast can react on actual weather changes for example.

$$\underline{f}_T = w \cdot \underline{f}_{T-1} + (1 - w) \cdot \frac{1}{n} \sum_{t=1}^{n} \frac{\underline{s}_t - \underline{s}_t^p}{\underline{s}_t^p} \tag{1}$$

Due to the modular bottom-up approach, these mathematical models could easily be replaced by other models. In the literature, neural networks as presented in [13,14] are very often used for grid state forecasts. Nevertheless, neural networks cannot calculate forecasts without measurements. To assure, that also in the case of absence of measurements, forecasts can be generated, the mathematical models should be kept as a backup solution, so that the plug-and-play functionality remains.

For the loads, this nodal treatment is not practicable, because single households cannot be simulated by a periodic mathematical model as well as renewable energy systems. The load behavior does not follow a single pattern, but rather shows some random effects with recurring patterns.

To get a first estimation, standard load profiles from [15] can be used for the load forecast, but if there are measuring sensors in the grid, these measurements should be used instead.

If every node in the grid is measured, an approach like for example the one shown in [16] can predict the loads all by their own. Especially in low voltage grids, this complete database will properly never be reached. Due to the fact that normally not every load is measured separately, the load is predicted for aggregated loads in grid areas as shown in Figure 2.

The load forecasts of these grid areas are based on the measurements of the past weeks, out of which an average load profile is built. If there are fewer measurements, the average of just one week or at least the last day is taken into account. By exponential smoothing, the power peaks, which are not every week at exactly the same time, are smoothed. In every time step, the load forecast is corrected by the actual measurements also by Equation (1).

Every section of this paper shows a new functionality of grid state forecasts to handle some specific new use cases. Other existing methods for grid state forecast do not handle specific use cases like these. A possible improvement of this load forecast will be presented in Section 2. In the end, the load has to be distributed to all nodes in the grid areas to get voltage and current values by a power flow calculation. For this distribution, several methods are possible, which will be shown and evaluated in Section 3. These distribution methods and the power flow calculation itself require lots of runtime. To examine if this effort is necessary, a second method without load distribution and power flow calculations has been implemented, which will also be evaluated in Section 4. In Section 5, it will be evaluated how an exchange of forecast information between different voltage levels can improve the grid state forecast quality. In Section 6, there is a short discussion of this concept, and in Section 7, a conclusion is made.

## 2. Improvement of the Load Forecast

The advantage of this modular-bottom-up approach consists in the possibility to react on different situations, which can occur during a longer time of cyclic executions of a grid state forecast software tool. One example can be the first start or a restart of the software tool. Grid state forecasts, which consist of neural networks or autoregressive models, cannot calculate valid forecasts in such a situation. They need a lot of training time, in which they collect measurements, before they are executable. By which new functionality the modular bottom-up approach can handle this situation, will be explained in the following.

At the beginning of recording measurements, there are not enough measurements to get a forecast out of them, for which at least a whole day is needed. To estimate the load base anyway, standard load profiles can be combined with the first measurements. For this method, the standard load profiles named H0 for households (written in the vector $\boldsymbol{h} := (h_1, \ldots, h_n)^T$), G0 for industry (written in the vector $\boldsymbol{g} := (g_1, \ldots, g_n)^T$), and L0 for agriculture (written in the vector $\boldsymbol{l} := (l_1, \ldots, l_n)^T$) are used [15]. These profiles are not suitable for single households, but for a few hundred in aggregation.

Due to this fact, this method will be working better in medium voltage grids than in low voltage grids, because there are more households in the grid areas, which are at large more similar to a standard load profile. Nevertheless, this method should be used also in low voltage levels, because this database is better than one single measured value, which does not identify the behavior over the whole day with sufficient accuracy. In the following, multiple possible cases with different available information of the loads are explained based on Figure 3.

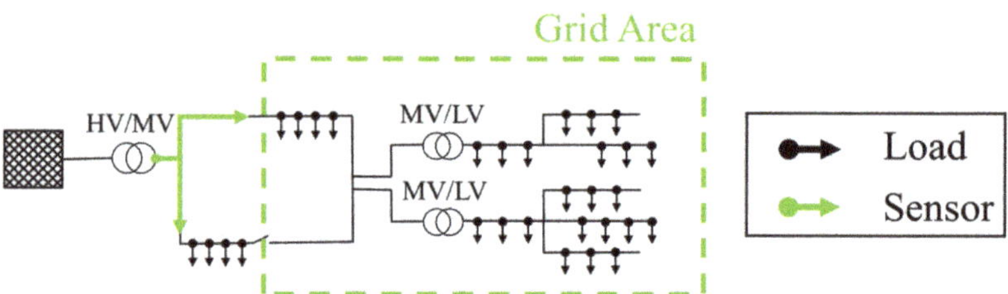


**Figure 3.** One example of a medium voltage grid area with different load types.

The most simple case consists in grid areas, where all loads and the number of load types ($N_h$, $N_g$,$N_l$) are known. This case exists if the considered grid is a LV grid and the loads in the grid area are known or in the case of a MV grid, where only MV/LV transformers with known load types can be estimated. According to Figure 3, this would be the case if the two MV/LV transformers would be the only loads in the highlighted grid area and the subordinate load types would be known. The calculation of the predicted grid area load $\boldsymbol{s}^p_{Load}$ in these cases is based on the Equation (3):

$$\boldsymbol{s}^p_{Load} = N_h \cdot \boldsymbol{h} + N_g \cdot \boldsymbol{g} + N_l \cdot \boldsymbol{l} \quad \text{with} \quad \boldsymbol{s}^p_{Load} := \left(s^p_{Load,1}, \ldots, s^p_{Load,n}\right)^T \tag{2}$$

At the system start or when the topology has switched and the old measurements cannot be used anymore, there is initially only one measurement. This data can be converted to the grid area power, which is not sufficient for calculating the grid area load forecast by an average day, but should still be used to calculate the number of loads in the grid area. For LV and MV grids, different cases can occur:

The first case is a medium voltage grid where all loads in the grid area are MV/LV transformers and the number of load types in the LV grids is not known or a LV grid, where the number of loads in the considered grid area is not known. In Figure 3, this would be the case if the two MV/LV transformers are the only loads in the highlighted grid area and the subordinate load types are not known. In this case, the assumption is made that all loads in the grid area are of the load type "household". The prediction of the grid area load by the calculation of the number of households with one measurement $\underline{s}_t$ of the time step $t$ is based on Equation (3).

$$\underline{s}_{Load}^{p^*} = \frac{\underline{s}_t}{h_t} \cdot \boldsymbol{h} \tag{3}$$

According to Figure 3, the opposite would occur if the two MV/LV transformers would not exist and all loads in one MV grid area would be industrial loads without available slave pointer values. However, if there are slave pointer values of the stations, these are assumed as constant loads. The first approximation in this case is using the G0 standard load profile to calculate the base load. This approximation is calculated using Equation (4).

$$\underline{s}_{Load}^{p} = \frac{\underline{s}_t}{g_t} \cdot \boldsymbol{g} \tag{4}$$

If there are $N_{MV/LV}$ MV/LV transformers and industrial loads in a MV grid area (as it is shown in Figure 3) and the number of load types in the LV grids is known, the grid area load is calculated using Equation (5).

$$\underline{s}_{Load}^{p} = \boldsymbol{x} + \frac{\underline{s}_t - x_t}{g_t} \cdot \boldsymbol{g} \text{ with } \quad \boldsymbol{x} := (x_1, \ldots, x_n)^T = N_h \cdot \boldsymbol{h} + N_g \cdot \boldsymbol{g} + N_l \cdot \boldsymbol{l} \tag{5}$$

If there are mixed loads and the number of load types in the subordinate grids is unknown, Equation (6) can be used.

$$\underline{s}_{Load}^{p} = \frac{\underline{s}_t}{h_t \cdot \left(N_g + N_{MV/LV}\right)} \cdot \boldsymbol{h} \cdot N_{MV/LV} + \frac{\underline{s}_t}{g_t \cdot \left(N_g + N_{MV/LV}\right)} \cdot \boldsymbol{g} \cdot N_g \tag{6}$$

An evaluation of this new method is unnecessary, because the alternative method would be to use the standard load profiles with an assumed specific number of households, which does not change. In this case, the forecast error depends on this assumption, so it can randomly match the real number of households but in most cases, it will not fit at all. Due to this, the results of a case study would just depend on the assumptions and would not be significant at all.

Other developed methods for short-term load forecasts, like [9,17], also depend on measurements and do not work at all in such cases. These methods would distribute no valid forecasts at all.

## 3. Improvement of the Grid Area Load Distribution

To calculate the voltage and current values at all nodes respectively branches, power flow calculations are used, but a power flow calculation cannot handle with grid area loads, because it needs a power value for every node in the grid. Therefore, the grid area loads have to be distributed to the nodes within the grid areas. To estimate the relation of the loads more realistically, some data like smart meter data, load types, or voltage and current measurements can be used. Different methods, how state estimation approaches can be adapted to a grid state forecast, will be presented in the following.

The use case for the methods, which will be shown in this chapter, is the following: At first, the grid area load is predicted by measurements of the past weeks. Second, the predicted load has to be distributed to the nodes within the grid area. For this distribution, different methods will be shown. For every grid area, it has to be decided which method is practicable for the load distribution. After the load distribution of every grid area, a power flow calculation can be done to calculate the voltage and current values for the considered grid.

### 3.1. *Distribution by Load Types or Smart Meter Data*

For distributing the predicted grid area loads to the nodes within the corresponding grid areas in low voltage grids, smart meter data are useful to calculate the household's percentage of the total grid area load for the considered time of the day, based on previous measurements. That previous smart meter measurements can improve the grid state estimation was already shown in [18]. This relation can be used for the distribution of the grid area load forecast.

At first, these calculated percentages of the total grid area load are assigned to the according smart meter nodes. After this, the remaining load is distributed to the other nodes depending on their load types and their relations to each other. One example of a load distribution is shown in Figure 4. In this case, the predicted grid area load is 20 kVA. The smart meter households get the calculated percentages depending on their previous smart meter data. The rest of the predicted power is distributed by the relations of the load types to the standard load profile of households. Due to this, the factor F for households is one. In this time step, the factor F of industry loads amounts 2. These factors are calculated with the predicted grid area power and are divided by the sum of all factors.

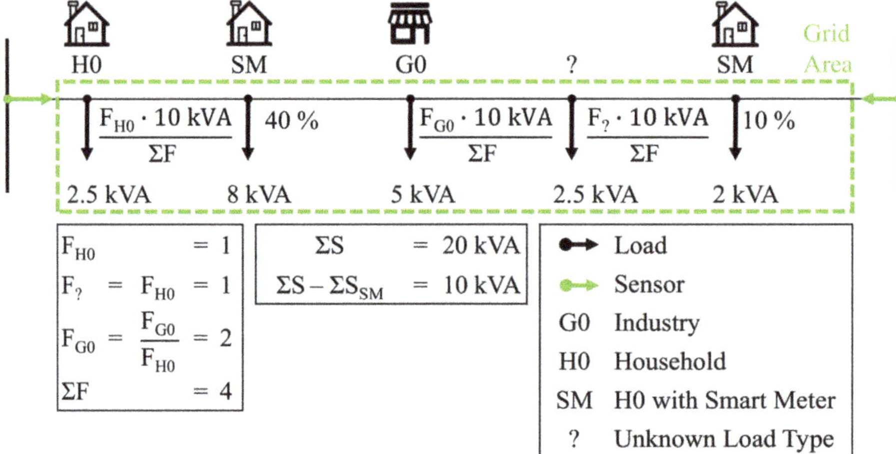


**Figure 4.** One example of a load distribution in a low voltage grid area at a time step with a relation between the G0 and the H0 standard load profiles of two.

These load types are the same as already presented in the section before. To use them for the load distribution, the relation of the standard load profiles must be established for each season, every weekday and every time. For example, on Sundays at noon, a household receives more energy than an industrial load—on weekdays it is the other way around.

If there is no information about the loads and no smart meter data, only a linear distribution is usable, so that the same load is applied to each load node.

### 3.2. *Improved Distribution by Measurements*

To distribute the grid area load to the nodes within, previous measurements are helpful to detect load centers, which obtain more load than the other nodes. The two corresponding methods, which will be explained in the following, are usable at every voltage level.

#### 3.2.1. Load Distribution by Voltage Measurements

If more than one voltage measurement exists in the grid area, which is the case especially in meshed grids, the voltage drop across the grid area can be examined, which can provide information about the load relation between the nodes in the grid area. This can be done by using the method to estimate the relation between the individual loads in the grid area presented in [19,20]. The novelty presented in the following is the usage of this method for a load distribution of a grid area load in the context of grid state forecasts by using previous measurements for the forecast of the load distribution in the future.

To detect the relation of the loads in one grid area by means of the voltage drop, the sensitivities in the grid area have to be investigated. The sensitivity describes the influence of a power change of one considered node on the voltage of the other nodes in the grid. How the sensitivity matrix can be calculated was already shown in [21]. The node with the highest sensitivity on the first sensor of the grid area is now detected as the best-case node, the node with the lowest sensitivity as the worst-case node. These designations result from the fact that the voltage drop would be small if the whole grid area load would be obtained by the best-case node and large if it would be obtained by the worst-case node. In addition, there is the linear case, where the whole load is shared equally among all load nodes in the grid area.

Which of these three cases is the most probable in the considered grid area at the considered time, can be calculated by evaluating the measurements of the weeks before, of which the measurements of the same time of the same weekday are considered. To proof, which case existed in the weeks before, a power flow calculation has to be done for every case. The measured load of the grid area is first linearly distributed, then placed completely on the worst-case node and finally on the best-case node. In all three variants, the remaining nodes in the grid are treated equally in all cases. The case whereby the calculated voltage values most closely match the measured values is chosen for this time step. Out of this calculation a factor $a$ for every week $k$ results, which can take values between one and three, whereby one stands for the worst-case, two stands for the linear distribution, and three represents the best-case. The weeks are weighted using Equation (7), so that the newest measurements are weighted the most to get the predicted factor $a^p$ of the load distribution of the considered time step.

$$a^p = \sum_{k=1}^{n} \frac{2 \cdot (n+1-k) \cdot a_k}{n \cdot (n+1)} \tag{7}$$

If there are more than two sensors in the grid area, then the two sensors are used for this procedure, which are separated by the most load nodes.

Figure 5 shows the used grid to simulate this use case of a grid state forecast. It is a real rural LV grid with 51 load nodes and 18 photovoltaic systems. It shows two highlighted grid areas, which are usable for the load distribution with the help of voltage measurements. The rest of the grid areas, which are not highlighted, are not usable for this method for the load distribution, because in each case there is just one single voltage sensor, which defines the grid area. In these areas, other methods like the linear load distribution or the distribution by smart meter data are usable.

To evaluate this method in the shown grid, a load flow simulation tool [22], which simulates the behavior of the load nodes by standard load profiles, is used instead of real grid measurements. For the generation of the photovoltaic systems, real measurements are used and converted to the installed powers, so that pseudo weather forecasts can be calculated by them. It runs a power flow calculation and creates pseudo measurements for the measured nodes, which are delivered to the grid state forecast. By using this simulation tool, the grid state forecast can be evaluated at all nodes in the grid and not just at the measured nodes. The process can be described as follows: At first, the load flow simulation tool delivers the pseudo measurements to the grid state forecast tool. After collecting enough measurements, a grid area load forecast can be done by calculating an average load profile out of the measurements of the previous days. This grid area load forecast is once distributed linearly, so that every load node gets the same load forecast, and once distributed by the measured voltage drops of the previous days. The nodal powers, which result by these two different approaches, are calculated to voltage and current values by a power flow calculation. By this method, the predicted voltages at the nodes between the measurements can be compared to the calculated values of the load flow simulation as reference.

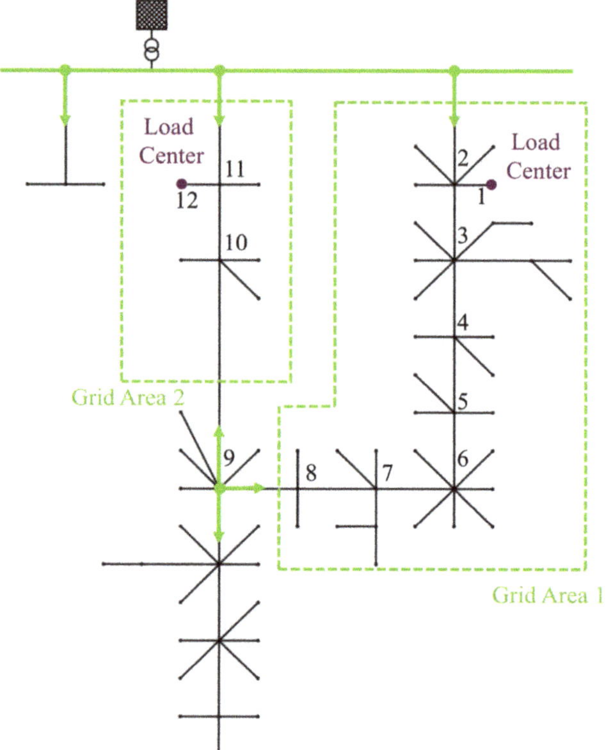


**Figure 5.** A low voltage grid with sensors at the transformer and the wiring closet at node 9.

To analyze the method to distribute the load within the grid areas, in every highlighted grid area is randomly placed one centered load, which obtains multiple power than the other nodes. These two load centers are set to a constant power demand through all time steps. Due to this, at all time steps the voltage-based load distribution should place the most of the predicted area load to the detected load center. Power losses are neglected.

Figure 6 shows the node voltages at one time step of the simulation. Through this method, the exact load center cannot be detected, but the voltage course over the nodes in general can be estimated much better with the help of older voltage measurements.

The analysis of the forecast errors are done by using the mean average percentage error (MAPE), because this measure is state of the literature and also used in like [8] or [10] for example. Equation (8) shows how the MAPE is calculated. $F_t$ stands for the predicted value (depends on the analysis, if it is voltage, current, or power) and $A_t$ stands for the real value of time step t. The MAPE is then the average of n time steps of this calculation.

$$\text{MAPE} = \frac{1}{n}\sum_{t=1}^{n}\left|\frac{A_t - F_t}{A_t}\right| \tag{8}$$

The MAPE over all nodes could be reduced from 0.69% to 0.62%, so this load distribution method is very advisable to generate more information out of the measurement database.

All voltage forecast errors of this grid areas over all simulated time steps are shown in Table 1. In general, it can be concluded, that the voltage-based load distribution can improve the grid state forecast a lot. To evaluate the applicability of this approach in the context of grid state forecasts in general, more simulations have to be done. Especially other topologies have to be analyzed, because the most LV grids are not meshed, like this example grid. Anyway, for this approach, meshed grids are nevertheless representative, because especially in meshed grids, the grid areas are defined by more than one sensor. Due to this fact, this approach will be used more often in meshed grids, than in grids with single strands.

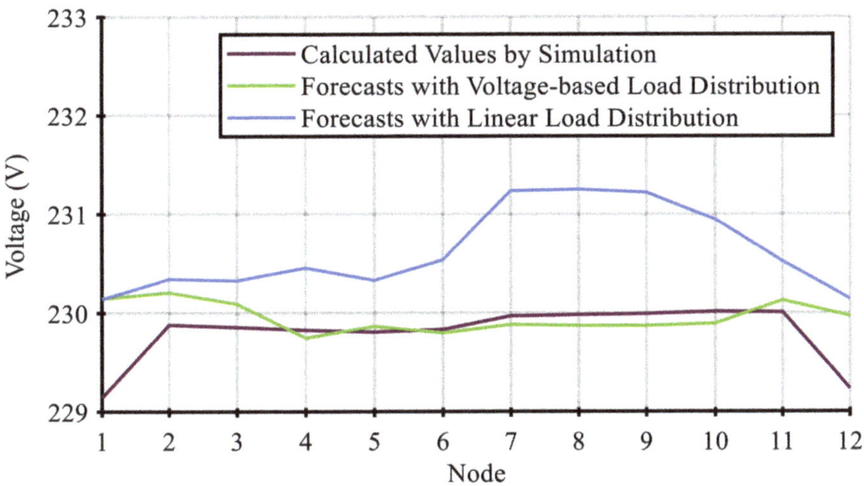


**Figure 6.** Improvement of the forecasts by using the voltage-based load distribution (one time step)

**Table 1.** Voltage forecast errors at the highlighted nodes.

| Node | 1 | 2 | 3 | 4 | 5 | 6 | 7 | 8 | 9 | 10 | 11 | 12 | Average |
|---|---|---|---|---|---|---|---|---|---|---|---|---|---|
| MAPE in % linear load distribution | 0.59 | 0.49 | 0.60 | 0.87 | 0.79 | 0.87 | 0.88 | 0.84 | 0.81 | 0.71 | 0.41 | 0.46 | 0.69 |
| MAPE in % voltage-based distribution | 0.60 | 0.48 | 0.55 | 0.74 | 0.68 | 0.74 | 0.78 | 0.75 | 0.72 | 0.63 | 0.38 | 0.37 | 0.62 |

3.2.2. Load Distribution by Current Measurements

In equivalent to the already shown method, current measurements can also be useful to estimate the load relation within one grid area. As a new expansion to the voltage depended procedure, a current-based method was developed, which is a new possibility to distribute the loads within one grid area. This method could be assumed for state estimations, too, if the following evaluation shows that it estimates the load distribution better than the linear distribution.

This method is useful, if there are not two voltage measurements limiting the grid area or if there are two measurements, but they are not at the edges of the grid area, so that they cannot give an useful information about the relation between all nodes in the grid area.

Mostly this concept is usable for ring structures, if there is just one voltage measurement, but two current measurements that define the considered grid area. In this case, the same procedure as in the case with two voltage measurements can be used. The past current measurements can give the information, if there exists mostly a best-case, a worst-case, or a linear case of the load relation in the grid area at the considered time step.

To proof the impact on the forecast quality, the already mentioned simulation tool can be used, too. The same grid is usable, but with another load scenario and another number of measurements. Power losses are neglected in this case, too. In this case, there are just the measurements at the transformer and not the measurements at the wiring closet, so that just one big grid area results, which is shown in Figure 7.

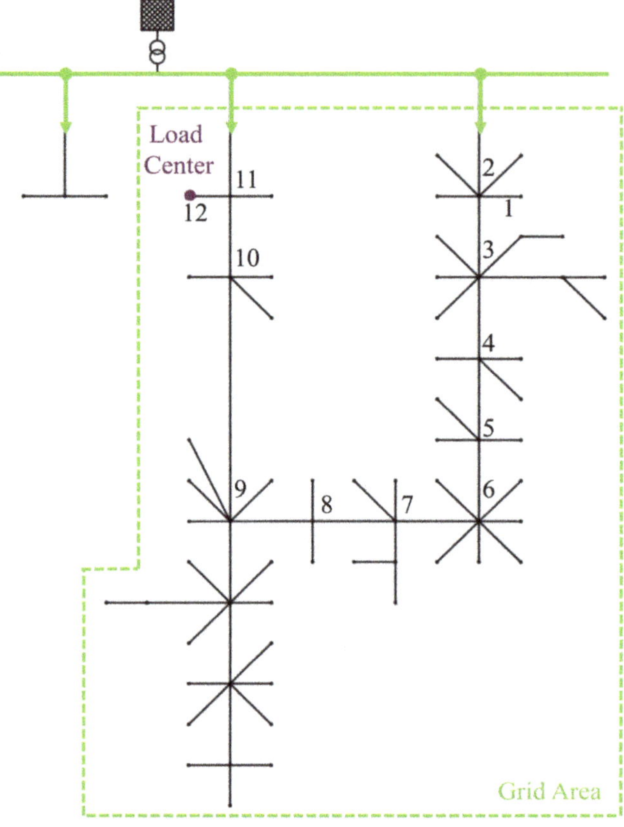


**Figure 7.** A low voltage grid with sensors at the transformer.

In the simulation, node twelve obtains much more load than the other nodes, to prove if the forecast detects this load center. All other time series are the same as in the example before. At node one, where was a load center too in the example before, a time series is set, which is more similar to the other nodes.

Due to that, the voltage-based method cannot be chosen and the load distribution by previous current measurement is executed by the grid state forecast.

Table 2 shows the mean average percentage errors of the highlighted nodes over all simulated time steps. The mean error of all node voltages is reduced from 0.55% to 0.44%, which is a big impact and underlines that this new method to distribute the grid area load is very useful.

One time step is shown in Figure 8. It is obvious that the load center is well detected and that the voltage forecasts at all nodes in this grid area can be improved by using the current measurements for the load distribution.

**Table 2.** Voltage forecast errors at the highlighted nodes.

| Node | 1 | 2 | 3 | 4 | 5 | 6 | 7 | 8 | 9 | 10 | 11 | 12 | Average |
|---|---|---|---|---|---|---|---|---|---|---|---|---|---|
| MAPE in linear load distribution | 0.39 | 0.39 | 0.47 | 0.69 | 0.63 | 0.69 | 0.7 | 0.67 | 0.64 | 0.54 | 0.30 | 0.47 | 0.55 |
| MAPE in current-based distribution | 0.25 | 0.25 | 0.31 | 0.50 | 0.43 | 0.50 | 0.59 | 0.59 | 0.59 | 0.53 | 0.38 | 0.38 | 0.44 |

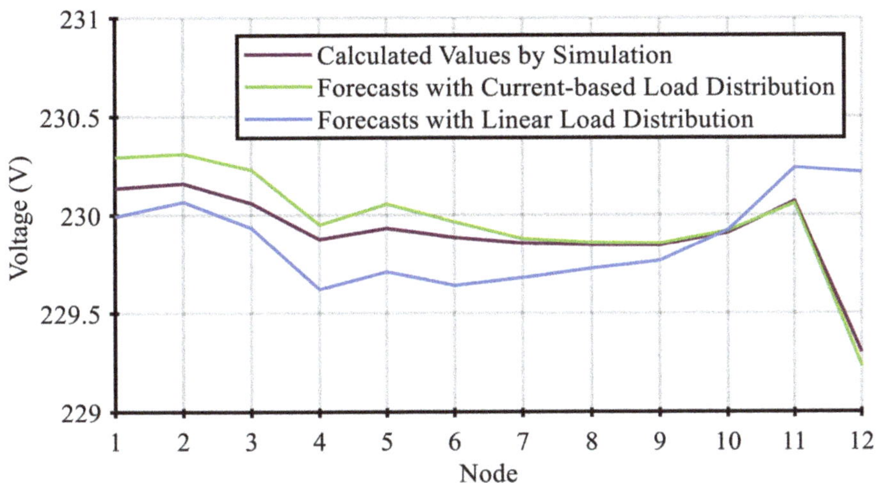


**Figure 8.** Improvement of the forecasts by using the current-based load distribution (one time step).

Figure 9 shows, that also the branch current forecasts can be improved by the presented method. In general, the mean current forecast error could be reduced by 5 %. This load center detection can thereby help to predict cable overloads. Particularly in regards to the increasing number of electric vehicles in the LV grids, this load center detection can be helpful to predict recurring charging patterns.

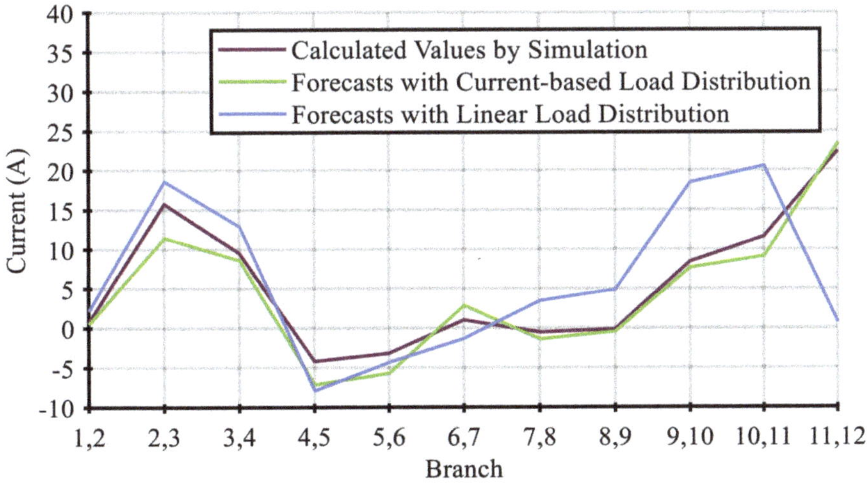


**Figure 9.** Simulated branch currents in comparison to the forecasts at one time step.

Finally, this approach is useful in the context of grid state forecasts, so it is conjecturable that it is also useful for state estimation in real time smart grids solutions. This application should be investigated in further research in the context of smart grids.

## 4. Simplification of the Grid State Estimation

The recently shown load distribution is needed for the power flow calculation, which is used to calculate the nodal voltages and the branch currents. Now it will be analyzed, if this big effort and the thereby resulting longer runtime can be reduced by using another method to calculate the voltage and current values without a power flow algorithm. For this scope, sensitivities could be helpful. This approach only calculates voltage and current forecasts for the measured nodes and branches. If the positions of the sensors are chosen appropriate and cover the important hotspots of the grid, these forecasts could suffice to classify the whole grid state. One advantage of the chosen sensitivity approach shown in [21] is that it is a deterministic method and always calculates a result. The power flow calculation instead does sometimes not converge. That the sensitivity approach can be useful in similar cases was already shown in [20].

The sensitivity matrix $\underline{\mathbf{X}}$ of the grid consists of the impact of apparent power changes on the changes of the voltages at all nodes.

Equation (9) shows the predicted voltage change at node $i$ $\Delta\underline{U}_i^p$ caused by a predicted change of the apparent power at node $k$ $\Delta\underline{S}_k^p$ based on the predicted voltage at node $k$ $\underline{U}_k^p$ (based on previous measurements as first estimation). The predicted apparent power change is the difference of the measured apparent power and the predicted value for the considered time step. This calculation can be done for every of the $N$ measured nodes.

$$\Delta\underline{U}_i^p = \sum_{k=1}^{N} \frac{\left(\Delta\underline{S}_k^p\right)^* \cdot \underline{X}_{k,i}}{3 \cdot \underline{U}_k^p} \tag{9}$$

The power changes at the $N$ measured nodes cause also current changes at the sensors in the grid. The number of the current sensors must not be the same as the number of voltage sensors. The predicted changes of the branch currents (for example $\Delta\underline{I}_{i,j}^p$ for the branch between the nodes $i$ and $j$) can be calculated equivalently considering Equation (10) and by using the impedance matrix $\underline{\mathbf{Z}}$.

$$\Delta\underline{I}_{i,j}^p = \sum_{k=1}^{N} \frac{\left(\Delta\underline{S}_k^p\right)^* \cdot \left(\underline{X}_{k,i} - \underline{X}_{k,j}\right)}{3 \cdot \underline{U}_k^p \cdot \underline{Z}_{i,j}} \tag{10}$$

Due to the fact that the nodal voltage $\underline{U}_k^p$ is needed for these Equations, this is just practicable for measured nodes and not for the nodes included in the grid areas. To consider also the power of $M$ grid areas, another sensitivity matrix $\underline{\mathbf{W}}$ is built, which describes the relation between the grid areas and all nodes in the grid. In this case, the variable $\Delta\underline{S}_k^p$ describes the predicted power difference of grid area $k$ and $\underline{W}_{k,i}$ describes the sensitivity of grid area $k$ on node $i$. The needed voltage value $\underline{U}_k^p$ is estimated by means of the sensor measurements, which limits the considered grid area. Thereby the Equations (11) and (12) result to consider all grid areas.

$$\Delta\underline{U}_i^p = \sum_{k=1}^{M} \frac{\left(\Delta\underline{S}_k^p\right)^* \cdot \underline{W}_{k,i}}{3 \cdot \underline{U}_k^p} \tag{11}$$

$$\Delta\underline{I}_{i,j}^p = \sum_{k=1}^{M} \frac{\left(\Delta\underline{S}_k^p\right)^* \cdot \left(\underline{W}_{k,i} - \underline{W}_{k,j}\right)}{3 \cdot \underline{U}_k^p \cdot \underline{Z}_{i,j}} \tag{12}$$

Finally, all of these voltage and current changes were calculated and added to the present measured values ($\underline{U}_i$ and $\underline{I}_{i,j}$). Thereby the forecasts for all sensors in the grid result according to the Equations (13) and (14).

$$\underline{U}_i^p = \underline{U}_i + \Delta\underline{U}_i^p + \Delta\underline{U}_{Slack}^p \tag{13}$$

$$\underline{I}_{i,j}^p = \underline{I}_{i,j} + \Delta\underline{I}_{i,j}^p \tag{14}$$

In Equation (13) $\Delta\underline{U}_{Slack}^p$ is the difference between the actual slack voltage at the transformer and the predicted slack voltage for the considered time step. This offset is important, because in the meantime of the forecast the situation in the higher voltage level could change and cause a non-negligible voltage change, which has an impact on all nodes in the considered grid. If the measured values of less than one day exists, the nominal voltage is assumed. Otherwise, the voltage value of the last day is used. If there are more than seven days, the voltage curve of the last same weekday is assumed. Finally, the difference between this forecast and the last measured values is taken into account in order to bring the forecasts closer to the current curve. This difference is considered with linearly decreasing weighting, so that it just has an effect on the next few time steps, because later time steps do not depend on actual short changes.

Figure 10 shows the results of executing the grid state forecast once with the sensitivity approach and once with the load distribution and power flow calculations for one grid. The used data are real measurements at 40% of the load nodes in the grid, for example at the transformer and at some wiring closets. These real measurements of one week for every season were evaluated.

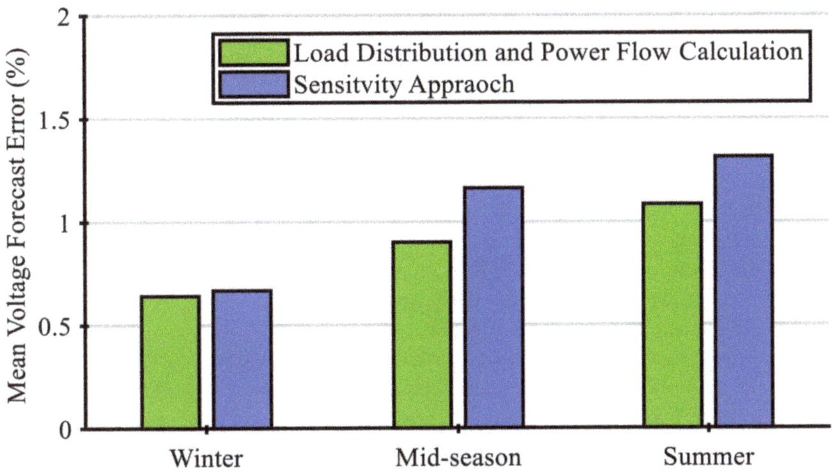


**Figure 10.** Evaluation of the error increase in one LV grid by using the sensitivity approach

The sensitivity approach is 8% faster (from 7.0 seconds to 6.6 seconds per time step with an Intel i7 quad-core processor and a RAM of 16 GB), but shows voltage errors (MAPE), which are 0.17% higher. In the context that the mean voltage error (MAPE) is 0.87% with the power flow calculation, this difference can be critical. In some situations, the sensitivity approach though can be better than the load distribution with the power flow, because the sensitivity approach uses the currently measured voltages to calculate the forecasts. For the power flow, only the slack voltage is taken into account and the other voltages (at unmeasured nodes and at measured nodes, too) are calculated by the power forecasts. If there are very small power changes predicted, the voltages do not change a lot. The sensitivity approach detects this behavior better, because it predicts an almost equal value to the current value. This could be the cause for the small difference between the two methods in winter. In winter, there are smaller power changes than in summer, because this grid is characterized by a high photovoltaic penetration. Due to this fact, more evaluations with more other grids have to be done to estimate the effect of the sensitivity approach on the forecast quality.

## 5. Exchanging Forecasts between Voltage Levels

As already mentioned, voltage changes at the slack node of the grid have a great impact on all nodal voltages. Due to this fact, the quality of the power forecasts is not the only important key factor in this use case. A good forecast of the temporal development of the slack voltage is even more decisive. Out of the considered grid, the forecast of the grid state of the higher voltage level is not feasible without knowing the grid parameters. Therefore, the only information, which is usable, are the previous measurements. It should be an aim to install a grid state forecast also in the higher voltage level, so that both forecasts can profit from each other, as is shown in Figure 11.

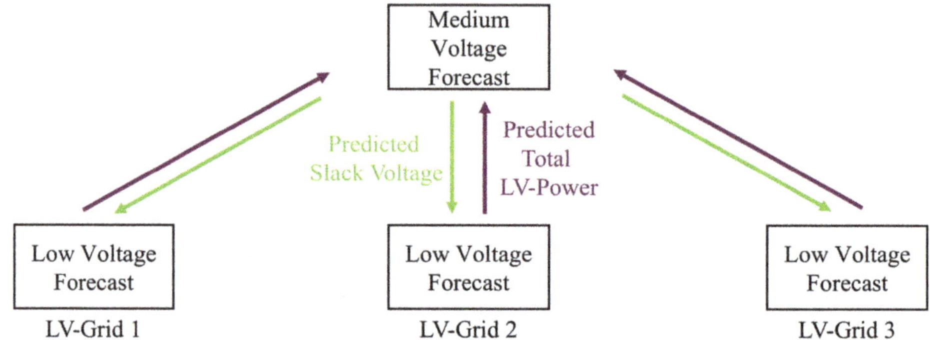


**Figure 11.** Procedure of exchanging information between the forecasts of two voltage levels.

At first, the LV forecast calculates voltage and power values for all nodes as well as current values for all branches. The MV forecast gets the predicted total power of the LV grid through a defined interface and handles it like a schedule for the according node. Then the MV forecast is executed. The predicted node voltages as output of this forecast can be used by the LV forecast as predicted slack voltage. This information can improve the forecast a lot, because without the predicted slack voltage as input, it has to be predicted without information from the higher voltage level and is very prone to error, which extends to all other nodes in the grid.

Figure 12 shows the measurements for one voltage sensor in one grid. The green line demonstrates the forecasts with known slack voltage. In reality, there would be also a forecast error out of the higher voltage level, but this is neglected in this case. Drawn in blue is the forecast without knowing the slack voltage. The shown forecasts were made at the simulation time step at 11:00 PM for the next 24 h. The figure shows the big effect of knowing the slack voltage for predicting the whole grid state. Through the exchange of forecasts between the different voltage levels, for example the planned step range at the transformer at 12:00 AM can be taken into account for the forecasts, which improves the forecasts very much.

The figure shows a node that is much meshed, so this effect does not just show up at transformer nearer nodes, but also at nodes, which are influenced also by the power changes at many other nodes.

Several of these analyses were made for this LV grid. The evaluations are based on one week for every season (summer, winter, and mid-season). Every half an hour, a forecast in a 5-min resolution for the next 24 h is made. Overall, the voltage forecast errors (MAPE) at the measured nodes improve generally by 0.19%, whereby the general voltage error (MAPE) amounts to 1.3% without known slack voltage. This demonstrates that the knowledge of the predicted slack voltage is decisive for a good forecast.

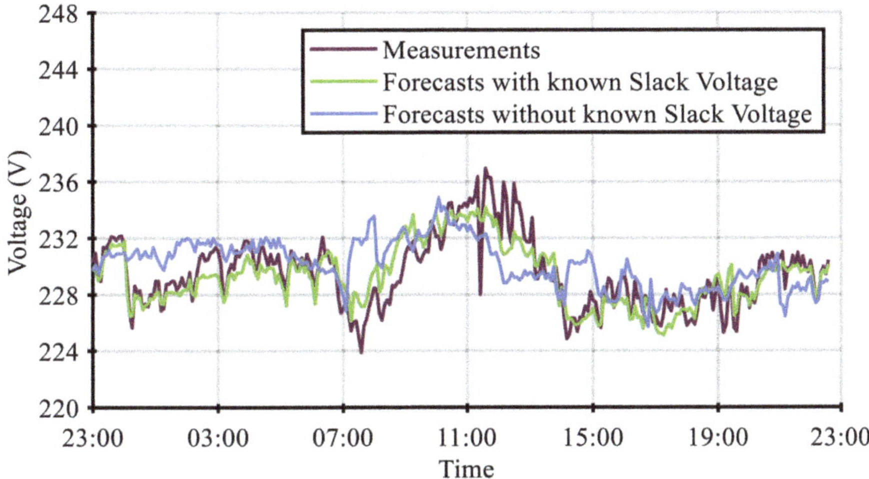


**Figure 12.** The effect of knowing the slack voltage on a node voltage forecast.

## 6. Discussion

In comparison to other approaches of grid state forecasts, the presented software application includes a few more use cases, as for example the missing of previous measurements. Other approaches, like they are shown in [6,7,23], obligatorily need measurements to calculate future grid states. This approach allows it to handle new situations in the grid without the necessity to restart the whole system. For example, this would be the case with a neural network, because it would have to train the system once again when a new generation system is integrated or the topology is changed. In addition, neural networks rely on a big database and a long executing time, which was already analyzed in [14].

The presented bottom-up approach instead needs less data (as shown in Section 2) and does not have to be trained at first. Due to these facts, a substitution of the shown grid state forecast with all its functionalities by a neural network is not possible, but some single forecast modules could be supplemented by other forecast models, like neural networks for example. Therefore, the possibility is given to combine this framework with other forecast methods.

The modular bottom-up approach allows changing single parts like the photovoltaic power forecast for example. The innovation consists therefore of the framework around all these single modules, which decides which module has to be used depending on the actual database.

The presented grid state forecast is realized in a standalone application and is very robust against external effects. In addition, it allows the exchange of forecast information between different voltage levels.

## 7. Conclusions

Different new functions for a grid state forecast were presented in this contribution. Some additional functionalities can improve the runtime of the software, some functionalities can improve the forecast quality:

- A new approach to calculate load forecasts completely without or with just a few measurements was shown. Thereby the base load can be calculated by the relation of standard load profiles and the measured value. This is a big advantage in comparison to other methods for load forecasts in the literature, because these need many measurements to calculate values for future time steps.
- It was shown that the load distribution by voltage or by current measurements, which is a new method to detect load centers in the grid, are better than a linear distribution. The current based

distribution can be useful for predicting the loads of electric vehicles, if the owner shows recurring charging patterns.

- It was shown that sensitivity approaches could be used for the state estimation instead of a load distribution and a power flow, but just if all critical nodes are measured. This is a condition, because the developed formulas are just usable for the forecast of measured nodes. At these nodes, the forecast quality is almost as good as the approach with the load distribution and the power flow.
- The most important presented functionality is the use of forecasts in coordination with other voltage levels. By the information exchange between the low voltage grids and the medium voltage grids, the slack voltages in the low voltage grids can be predicted much more precisely. It was shown that the knowledge of the slack voltage has a significant impact on all nodes in the analyzed grid and could improve the forecast quality in all cases.

In the future, the grid state forecast can be improved by even more functionalities. According to the modular bottom-up approach, new models can be integrated easily.

**Author Contributions:** Conceptualization, J.H.; methodology, J.H., M.M., K.K. (Kamil Korotkiewicz), F.P.; software, J.H.; validation, J.H.; writing—original draft preparation, J.H.; writing—review and editing, M.M., K.K. (Kamil Korotkiewicz), F.P., K.K. (Kevin Kotthaus), S.P., M.Z. All authors have read and agreed to the published version of the manuscript.

**Funding:** The research for this paper was sponsored by the German Federal Ministry for Economic Affairs and Energy—Project Flex2Market (03ET4043A)—The authors take responsibility for the published results.

**Conflicts of Interest:** The authors declare no conflict of interest.

## References

1. Ram, M.; Bogdanov, D.; Aghahosseini, A.; Gulagi, A.; Oyewo, A.S.; Child, M.; Caldera, U.; Sadovskaia, K.; Farfan, J.; Barbosa, L.; et al. *Global Energy System based on 100% Renewable Energy –EnergyTransition in Europe Across Power, Heat, Transport and Desalination Sectors*; LUT University: Lappeenranta, Finland; Energy Watch Group: Berlin, Germany, 2018.
2. Kotthaus, K.; Hermanns, J.; Paulat, F.; Pack, S.; Meese, J.; Zdrallek, M.; Neusel-Lange, N.; Schweiger, F.; Schweiger, R. Concrete design of local flexibility markets using the traffic light approach. In Proceedings of the CIRED Workshop 2018 on Microgrids and Local Energy Communities, CIRED Workshop, Ljubljana, Slovenia, 7–8 June 2018.
3. Hermanns, J.; Pack, S.; Kotthaus, K.; Zdrallek, M.; Schweiger, F.; Schweiger, R.; Raczka, S.; Baumeister, C. Preparation of a field test to evaluate a local flexibility market as a smart grid add-on. In Proceedings of the 2019 Conference on Sustainable Energy Supply and Energy Storage Systems, IEEE PES NEIS Conference, Hamburg, Germany, 19–20 September 2019.
4. Biegel, B.; Andersen, P.; Stoustrup, J.; Rasmussen, K.S.; Hansen, L.H.; Ostberg, S.; Cajar, P.; Knudsen, H. The value of flexibility in the distribution grid. In Proceedings of the 2014 IEEE PES Innovative Smart Grid Technologies Europe, IEEE PES ISGT-Europe, Istanbul, Turkey, 12–15 October 2014.
5. Hayes, B.P.; Prodanovic, M. State Forecasting and Operational Planning for Distribution Network Energy Management Systems. *IEEE Trans. Smart Grid* **2016**, *7*, 1002–1011. [CrossRef]
6. Antoncic, M.; Ilkovski, M.; Blazic, B. State forecasting in distribution networks. In Proceedings of the 2019 IEEE PES Innovative Smart Grid Technologies Europe (ISGT-Europe), Bucharest, Romania, 29 September–2 October 2019.
7. Zhao, J.; Zhang, G.; Dong, Z.Y.; La Scala, M. Robust Forecasting Aided Power System State Estimation Considering State Correlations. *IEEE Trans. Smart Grid* **2018**, *9*, 2658–2666. [CrossRef]
8. Dehalwar, V.; Kalam, A.; Kolhe, M.L.; Zayegh, A. Electricity load forecasting for Urban area using weather forecast information. In Proceedings of the 2016 IEEE International Conference on Power and Renewable Energy (ICPRE), Shanghai, China, 21–23 October 2016; pp. 355–359.
9. Bennett, C.; Stewart, R.; Lu, J. Autoregressive with Exogenous Variables and Neural Network Short-Term Load Forecast Models for Residential Low Voltage Distribution Networks. *Energies* **2014**, *77*, 2938–2960. [CrossRef]

10. Rejc, M.; Einfalt, A.; Gawron-Deutsch, T. Short-term aggregated load and distributed generation forecast using fuzzy grouping approach. In Proceedings of the 2015 International Symposium on Smart Electric Distribution Systems and Technologies (EDST), Vienna, Austria, 8–11 September 2015; pp. 212–217.
11. Neusel-Lange, N.; Oerter, C.; Zdrallek, M. State identification and automatic control of smart low voltage grids. In Proceedings of the 2012 3rd IEEE PES Innovative Smart Grid Technologies Europe (ISGT Europe), Berlin, Germany, 14–17 October 2012.
12. Kolev, V.; Sulakov, S. Short-term power output forecasting of the photovoltaics in Bulgaria. In Proceedings of the 2018 10th Electrical Engineering Faculty Conference (BulEF), Sozopol, Bulgariam, 11–14 September 2018; pp. 1–4.
13. Khana, I.; Zhua, H.; Yaoa, J.; Khana, D. Photovoltaic Power Forecasting based on Elman Neural Network Software Engineering Method. In Proceedings of the 2017 8th IEEE International Conference on Software Engineering and Service Science, ICSESS, Beijing, China, 20–22 November 2017.
14. Hamid Oudjana, S.; Hellal, A.; Hadj Mahamed, I. Short term photovoltaic power generation forecasting using neural network. In Proceedings of the 2012 11th International Conference on Environment and Electrical Engineering, Venice, Italy, 18–25 May 2012; pp. 706–711.
15. Schieferdecker, B.; Fünfgeld, C.; Meier, H.; Adam, T. Repraesentative VDEW-Lastprofile. In *VDEW-Materialien*; VDEW: Frankfurt, Germany, 1999.
16. Yuce, B.; Mourshed, M.; Rezgui, Y. A Smart Forecasting Approach to District Energy Management. *Energies* **2017**, *10*, 1073. [CrossRef]
17. Li, Y.; Guo, P.; Li, X. Short-Term Load Forecasting Based on the Analysis of User Electricity Behavior. *Algorithms* **2016**, *9*, 80. [CrossRef]
18. Samarakoon, K.; Wu, J.; Ekanayake, J.; Jenkins, N. Use of delayed smart meter measurements for distribution state estimation. In Proceedings of the 2011 IEEE Power and Energy Society General Meeting, 2011 IEEE Power & Energy Society General Meeting, San Diego, CA, USA, 24–29 July 2011.
19. Ludwig, M.; Korotkiewicz, K.; Dahlmann, B.; Zdrallek, M.; Derksen, C.; Loose, N.; Törsleff, S.; Wassermann, E. Agent-based grid automation in distribution grids: Experiences under real field conditions. In Proceedings of the CIRED Workshop 2018 on microgrids and local energy communities, CIRED Workshop, Ljubljana, Slovenia, 7–8 June 2018.
20. Steinbusch, P.; Modemann, M.; Wazifehdust, M.; Zdrallek, M. Fast Distribution Grid State Estimation Using Improved Sensitivity Analysis. In Proceedings of the 8th IEEE PES Innovative Smart Grid Technologies Conference Europe, IEEE PES ISGT-Europe, Sarajevo, Bosnia and Herzegovina, 21–25 October 2018.
21. Wolter, M. *Grid State Identification of Distribution Grids*; Shaker: Aachen, Germany, 2008.
22. Dorsemagen, F. *Zustandsidentifikation von Mittelspannungsnetzen für eine übergreifende Automatisierung der Mittel- und Niederspannungsebene*, 1st ed.; epubli: Berlin, Germany, 2018.
23. Hayes, B.P.; Gruber, J.K.; Prodanovic, M. A Closed-Loop State Estimation Tool for MV Network Monitoring and Operation. *IEEE Trans. Smart Grid* **2015**, *6*, 2116–2125. [CrossRef]

*Article*

# The Design of a Low Cost Phasor Measurement Unit

**Antonio Delle Femine, Daniele Gallo, Carmine Landi and Mario Luiso ***

Department of Engineering, University of Campania "Luigi Vanvitelli", 81031 Aversa (CE), Italy
* Correspondence: mario.luiso@unicampania.it; Tel.: +39-0815010484

Received: 12 June 2019; Accepted: 6 July 2019; Published: 10 July 2019

**Abstract:** The widespread diffusion of Phasor Measurement Units (PMUs) is a becoming a need for the development of the "smartness" of power systems. However, PMU with accuracy compliant to the standard Institute of Electrical and Electronics Engineers (IEEE) C37.118.1-2011 and its amendment IEEE Std C37.118.1a-2014 have typically costs that constitute a brake for their diffusion. Therefore, in this paper, the design of a low-cost implementation of a PMU is presented. The low cost approach is followed in the design of all the building blocks of the PMU. A key feature of the presented approach is that the data acquisition, data processing and data communication are integrated in a single low cost microcontroller. The synchronization is obtained using a simple external Global Positioning System receiver, which does not provide a disciplined clock. The synchronization of sampling frequency, and thus of the measurement, to the Universal Time Coordinated, is obtained by means of a suitable signal processing technique. For this implementation, the Interpolated Discrete Fourier Transform has been used as the synchrophasor estimation algorithm. A thorough metrological characterization of the realized prototype in different test conditions proposed by the standards, using a high performance PMU calibrator, is also shown.

**Keywords:** Power System Measurement; Phasor Measurement Unit; low cost; synchrophasor; voltage measurement; current measurement; synchronization

---

## 1. Introduction

The need for the best estimate of the power system's state is recognized to be a crucial element in improving its performance and its resilience to face catastrophic failures. Thus, one of the most important advancement expected from smart grid technology is the strengthening of the management of the power system [1]. Currently, most of the control actions of a power system are performed through an open-loop type centralized control that implements only steady-state security functions. This applies since the Wide Area Measurement Systems (WAMS) typically devoted to this aim have long latency time. This, obviously, places some limits in terms of the level of stability, reliability and safety of the supervised power system. For this reason, in recent years, synchrophasor technologies and the related monitoring devices called Phasor Measurement Units (PMUs) have received a lot of attention [2–5]. A PMU measures the instantaneous voltage, current, frequency and the Rate Of Change Of Frequency (ROCOF) at specific locations in an electric power transmission system; then, it converts the measured parameters into phasor values, typically with a rate of 25 or more, per second. Finally, it also adds a precise time stamp to these phasor values, turning them into synchrophasors. Time stamping allows these phasor values, provided by PMUs in different locations and across different power industry organizations, to be correlated and time-aligned and so properly combined. The resulting information enables transmission grid planners and operators to have a high-resolution "picture" of the conditions throughout the grid in real time [1]. Thus, with a large-scale implementation of WAMS using PMUs and Phasor Data Concentrators (PDCs) in a hierarchical structure, it become possible to perform a closed loop automatic monitoring and control of the power system to steer it away from transient or voltage instability, through corrective actions initiated during a state of emergency [4,5].

The number of installed PMUs worldwide is constantly increasing; nevertheless, the cost of these devices is still a source of concern for widespread installations, [6–8]. Thus, a certain effort in research field is devoted in developing methodology to better observe, understand and manage the grid but limiting the number of installed devices, [9–11]. This paper tries to face the same problem but from a different point of view: it proposes a design approach for implementing a PMU only adopting low cost hardware, thus making it possible to use them on a large scale.

In recent years, there have been significant research contributions on the improvement of the PMU performances adopting different algorithms [12–19], or on the PMU-based event detection [20,21]. Few works, to the best of the author's knowledge, focused on the implementation of this kind of instruments. In [22], an example of PMU for distribution grids is presented; however, the implementation details are not disclosed. More details are given in [23] where the described PMU prototype is based on a field programmable gate array with high performance. Despite this, the chosen hardware platform (i.e., a National Instrument Compact RIO) is quite expensive. In [24], a prototype of a PMU, based on a microcontroller, is presented; the synchronization is obtained through a GPS signal received from a Wireless Fidelity (Wi-Fi) module. No details are given on the synchronization, nor device characterization is presented. In [25], the OpenPMU project, an open platform for the development of PMU technology, is presented. No specific details on hardware implementation but just a rough cost, of about 1000 $, are given; moreover, nothing is said about the instrument performance. In [26], a development of an analog-to-digital converter (ADC) for PMU applications, with GPS synchronization, based on an open hardware development platform, is discussed. It makes use of external devices, such as a GPS receiver, a Phase Locked Loop (PLL) circuit and an ADC, managed by the powerful BeagleBone Black board. Some issues arise: the development board is not an industrial product, so not suitable for harsh environment like substations; moreover, the performance is not accurately evaluated in comparison with a reference instrument. In [27], a technique to lock the sampling frequency of an ADC, managed by a Digital Signal Processor (DSP), is discussed; it does not use a GPS disciplined oscillator, but a simple GPS receiver. It is based on a non-uniform sampling of the signal, since the sampling period is continuously varied between two discrete values. Good synchronization results are shown in the paper; however, the performance is evaluated in a very simple condition, which is a sine wave with constant frequency, amplitude and phase. In fact, the non-uniform sampling may introduce phase noise and worsen the accuracy of synchrophasor phase estimation.

It is worthwhile to emphasize that in all the cited papers, even if the performance is experimentally evaluated, merely rough experiments using signals of a few volt are executed, excluding input transducers. However, it is known that input transducers are typically the major source of uncertainty in measurement chains for power systems [28–37]; this issue, specifically for PMU application, is also demonstrated in [38].

In this paper, a design approach for a low-cost prototype of PMU, with a detailed description of its hardware and firmware implementation, is presented. In addition, a thorough metrological characterization of the realized prototype is shown: it has been performed using a metrological grade reference instrument, the Fluke 6135A/PMUCAL, using voltage and current levels typical of low voltage power systems, which are similar to those which can be found in primary, or secondary, substations, at the output of Voltage and Current Instrument Transformers (VT and CT). The paper is organized as follows. In Section 2 some basic recalls on PMU are given. Section 3 presents the hardware implementation of the proposed system, including the analog input adaption stage. Section 4 describes the firmware implementation, along with the techniques used to obtain the synchronization and to improve the measurement accuracy. Section 5 shows the metrological characterization of the prototype and, finally, Section 6 draws the conclusions.

## 2. Fundamentals of PMU

A phasor is a complex number that represents both the magnitude, $A$, and phase angle, $\varphi$, of the voltage or current sinusoidal waveforms pulsating with an angular frequency $\omega = 2\pi f$ (with $f$ equal to 50 Hz or 60 Hz in different power systems) at a specific point in time (shown in Figure 1).

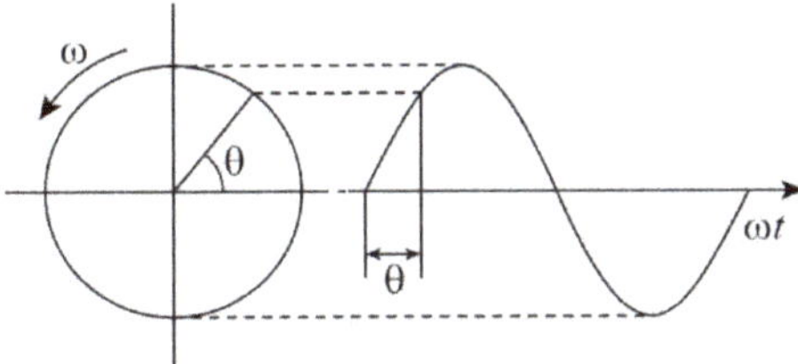


**Figure 1.** Synchrophasor representation.

PMUs measure root mean square (rms) amplitude, phase, frequency and ROCOF of both current and voltage and this collection of grid condition data, which are time-synchronized, is called phasor data. Every PMU measurement obtains a timestamp derived from the GPS universal time. When a phasor measurement is timestamped, it is called a synchrophasor. In this way, PMU measurements performed in different locations can be synchronized and time-aligned and, therefore, combined to provide a detailed view of a wide geographical area. This, moreover, help the system operators to maintain the healthiness of the network.

PMUs sample at speeds of up to 50 observations per second (or 60 in USA system), whereas conventional monitoring technologies (such as Supervisory Control And Data Acquisition, SCADA) measure once every two to four seconds. However, in order to allow the comparison of the electrical quantities of the nodes (amplitude and phase of the voltages and currents), the measurements must be made at common sampling instants. The absolute time reference can be used to synchronize the simultaneous sampling of voltage and current signals. The standards [2,3] define the reference time instants in which the PMU must measure the electrical signals and the levels of accuracy that equipment should meet for the various classes of accuracy. In order to face these requirements several test conditions and performance verifications are prescribed. A crucial role in performance verification is played by the synchronization stage: in fact, as it is stated in [2,3], a synchronization error of 1 μs results itself in a Total Vector Error (TVE) of 1 %. All the recalled requirements reflect on the design of a PMU measurement system. A basic architecture, typically adopted for PMU implementation, is reported in Figure 2.

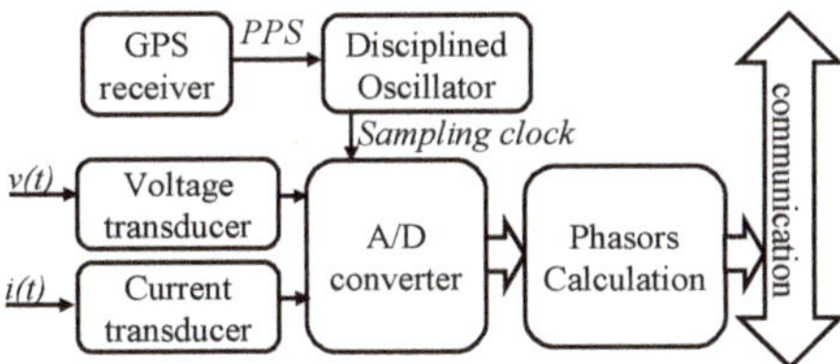


**Figure 2.** Basic architecture of a Phasor Measurement Unit, which typically include a GPS Disciplined Oscillator.

The core of this architecture is the oscillator disciplined by the Pulse Per Second (PPS) signal coming from GPS receiver, which can give a sampling clock accurately synchronized to the absolute time reference, i.e., the Universal Time Coordinated (UTC). In this way, the synchronization requirements are satisfied as the analog signals are sampled synchronously with the absolute time reference.

Nevertheless, the GPS Disciplined Oscillator (GPSDO) is not a cheap component and, to obtain a low-cost implementation, this component should be removed.

In addition, as is demonstrated in [38], the input transducers and the analog signal conditioning stages could be the major source of uncertainty in PMU measurement systems. Therefore, particular attention should be also paid to the design and usage of these components.

## 3. Hardware Implementation

### 3.1. System Architecture

In order to obtain an adequate level of accuracy and, at the same time, keep low the hardware cost, reference is made to the architecture reported in Figure 3 and only cheap components have been chosen. The input stages are constituted by low cost voltage and current transducers equipped with suitable analog conditioning stages to adapt the signal level to the input range of the ADC. The core of the instrument is a low-cost Advanced Reduced Instruction Set Computer (RISC) Machine (ARM) microcontroller unit (MCU) with integrated ADC (analog-to-digital converter) and Ethernet interface. It is responsible for the absolute time synchronization, data acquisition, signal processing and data communication. A key feature of the design is the lack of a GPSDO. Instead, a simple GPS receiver is used and all the synchronization is derived by the PPS signal. Therefore, a suitable signal processing is adopted to obtain measurements synchronized to the UTC, as it is better explained in Section 4. Voltage and current synchrophasors, frequency and ROCOF are obtained by processing the synchronized signal samples through an Interpolated Discrete Fourier Transform (IpDFT) algorithm [14]. Measurement data are communicated through a Transmission Control Protocol (TCP) socket to a host Personal Computer (PC); for the scope of this work, the standard Institute of Electrical and Electronics Engineers (IEEE) Std C37.118.2-2011 [39] has not been considered.

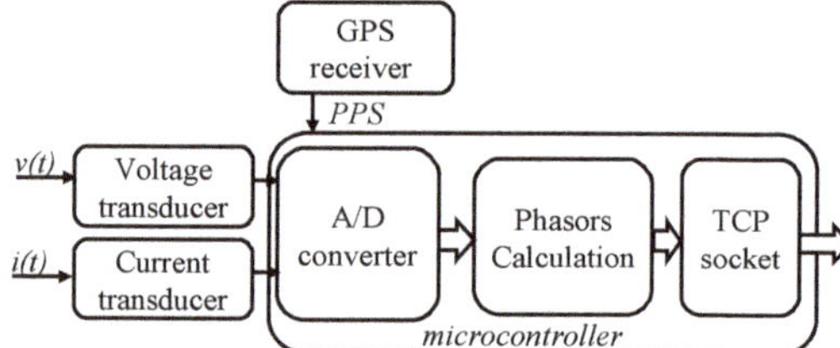


**Figure 3.** The proposed architecture for a low-cost Phasor Measurement Units (PMU).

### 3.2. Input Stage

The used voltage and current transducers are the LEM LV 25-P and the LEM LA 25-NP, respectively. Their conditioning circuits have been designed as simple as possible, using the lowest number of active components as possible, in order to keep, at the same time, the cost low and the signal-to-noise ratio as high as possible. The conditioning circuit for the voltage transducer is shown in Figure 4, where T1 represents the transducer. It converts a rated rms input current of 10 mA in a current of 25 mA, with a maximum rms input voltage of 700 V. The input signal range has been considered limited to a rms value of 300 V so the input resistance $R_1$ is chosen equal to 30 k$\Omega$ obtaining an output bipolar current of 25 mA. Then, in order to obtain a peak-to-peak unipolar voltage output of about 3.3 V (i.e., the input range of the microcontroller ADC) a resistor $R_2 = 46\ \Omega$ is inserted in series and an offset voltage of 1.65 V, directly derived from ADC reference voltage, is added as shown in Figure 4. The values of the other components are $R_3 = 10$ k$\Omega$ (in order to drain a maximum current of about 170 µA from the microcontroller) and $C_1 = 47$ nF (in order to cut high frequency noise). The adopted current conditioning circuit is very similar to that previously presented for voltage, the only difference in the scheme is that current transducer has no need of input resistance.

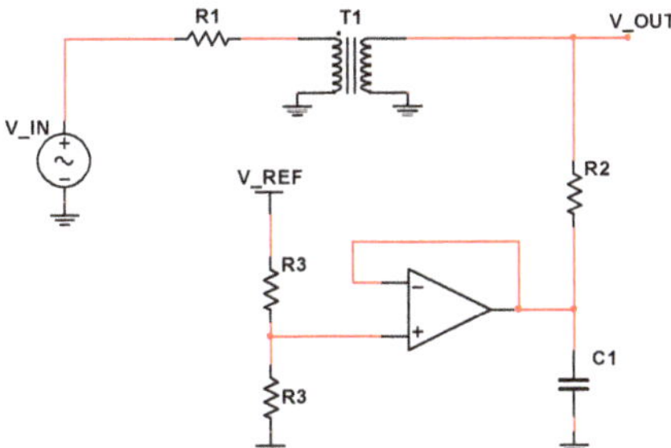


**Figure 4.** Simplified circuit topology of the voltage channel conditioning stage.

### 3.3. Microcontroller

The signals obtained from input stages are directly connected to two analog inputs of a STM32F407V MCU. This microcontroller is based on an ARM Cortex-M4 32 bit core with Floating Point Unit (FPU), it reaches 210 Dhrystone Mega Instructions Per Second (DMIPS) running at 168 MHz clock. It carries 512 kB of Flash and 192 kB of Static Random Access Memory (SRAM). It integrates several peripheral: Universal Serial Bus On-The-Go (USB OTG) HS/FS, Ethernet, 17 timers, three ADCs, 15 communication interfaces and camera interface. For this project, different features of MCU are used: two different 12-bit ADCs to acquire input signals, two internal 16-bit timers running at 168 MHz for timing and synchronization management, a Universal Asynchronous Receiver Transmitter (UART) to communicate with GPS receiver, an external input interrupt to receive the PPS signal and the Ethernet Media Access Control (MAC) interface to implement communication.

### 3.4. GPS Receiver

It is worthwhile to underline that, for cost reasons, only simple GPS receivers without disciplined oscillator output have been considered and an external Original Equipment Manufacturer (OEM) GPS receiver module Linx-rxm-GPS-FM has been chosen. It is a self-contained receiver based on the MediaTek MT3339 chipset, it can simultaneously acquire on 66 channels and track on up to 22 channels. This gives the module fast lock times even at low signal levels. The module outputs standard National Marine Electronics Association (NMEA) data messages through a UART interface.

## 4. Firmware Implementation

The firmware is fully developed in C language; Keil Microcontroller Development Kit (MDK) ARM development environment has been chosen for its high performance toolchain. The presented work has been developed on a low cost microcontroller, with relatively poor hardware features; consequently, great challenges are constituted by the firmware optimization and the accurate metrological characterization to evaluate and compensate the systematic errors. In particular, the main challenges have been related to: (1) the lack of specific synchronization hardware, (2) the lack of high resolution 32-bit timer at 168 MHz, (3) the poor Digital Signal Processing (DSP) performance of the floating-point unit (it does not implement 64 bit double-precision calculations) and (4) the low RAM size. In this section, some details on the main parts of firmware will be described, along with the techniques adopted to overcome the hardware limitations.

### 4.1. Synchronization

A PMU shall be capable to receive time synchronization from a reliable and accurate source, such as the GPS, and to perform phasor measurements synchronized to UTC time, with accuracy sufficient to meet the requirements specified in the standards [2,3]. It should perform all the measurements and report the results at a constant reporting rate, expressed in terms of frames per second (fps). The reporting rate is an integer number (i.e., 50/60 fps, 25/30 fps, 10/12 fps, etc.) and it defines the reporting time instants at which the PMU shall report the measurement results. For a reporting rate N fps, the N reporting times are evenly spaced through each second with first reporting time coincident with the UTC second rollover (e.g., coincident with a 1 PPS provided by GPS, see Figure 5). The typical solution adopted by commercial PMUs is the use of a 10 MHz GPSDO. Starting from this system clock, a sampling clock is derived, so that also sampling instants are referenced to UTC and the constraint of evenly spaced measurements is simply obtained by performing analysis on a constant number of samples (see Figure 6). In this paper, with the aim of cost saving, a different approach, based on a cheap GPS receiver and on a complex firmware technique based on numeric synchronization and resampling, is proposed. In the presented prototype, the internal MCU clock is not disciplined and, thus, it is not aligned to absolute time (see Figure 7). Moreover, it is obtained by multiplying the output frequency of a low cost quartz crystal through the PLL circuit (integrated in the MCU) and so affected by frequency jitter and thermal drift; its frequency (168 MHz), however, is much higher than the typical GPSDO frequency. Therefore, the system clock and internal timers are used to build a reference time-base (TB) and a firmware procedure manages the timers to keep their rollover periodicity locked to PPS and thus to UTC absolute time. To this aim, an internal counter (Time-Base Counter, TBC), running at 168 MHz, is used and configured with a counting number that is continuously estimated and corrected to keep its periodicity as close as possible to an integer fraction of 1 s (i.e., 1 s/50 = 20 ms). In fact, at each PPS event coming from the GPS receiver, a specific Interrupt Service Routine (ISR) calculates how many system ticks have been elapsed from the previous PPS event. This number is used to correct the counting number adopted in the next second to produce a counter rollover each 20 ms: the number of ticks, divided by 50, is the fractional number of ticks that corresponds to 20 ms and it defines the reporting time at 50 fps, the maximum reporting rate considered (i.e., related to 50 Hz power frequency). Obviously, since the number of ticks, divided by 50, can be a decimal number, whereas the counter accepts only integer numbers, a specific management strategy, explained in Section 4.3, has been used. Other reporting rates can be obtained by decimation. The correction is calculated adopting a discrete Proportional Integrative Derivative (PID) control algorithm, as better explained in Section 4.2. In this way, since the counter rollover is precisely produced every 20 ms (i.e., 50 pulses per second, 50-pps), it can be used to obtain the synchronized reporting time instants. However, in order to have synchronized measurements of the phasor, having a synchronized TB is not enough. In fact, it is necessary to sample the input signals synchronously with the absolute time reference, which is not the case at hand. Therefore, at each TB rollover, MCU takes and stores, in a dedicated queue, the timing information needed to perform a signal resampling, that is the number of ticks elapsed from last sampling time, called control info (CI). In fact, with CI, the actual synchronized sampling instants are estimated and then the acquired data are resampled with linear interpolation, in order to have a signal synchronously sampled. The data acquired between two subsequent TB events is called frame. The subsequent processing stage manipulates the frames and extracts the phasors.

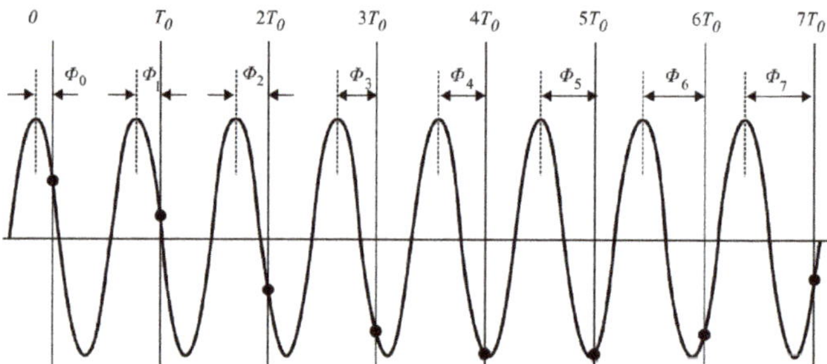


**Figure 5.** A sinusoid with a frequency f, after the Pulse Per Second (PPS), is observed with a reporting time of $T_0$ seconds.

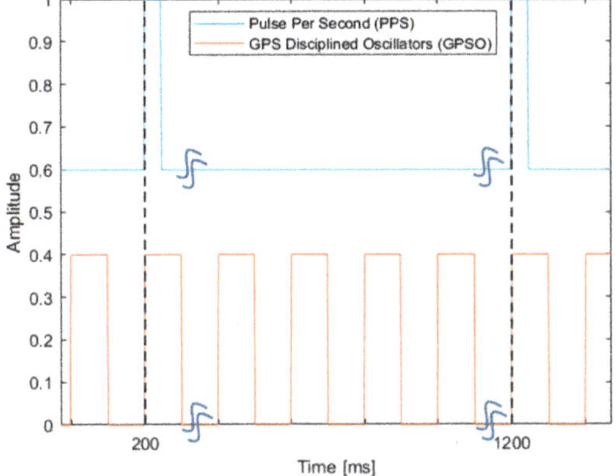


**Figure 6.** Example of a GPS Disciplined Oscillator (GPSDO).

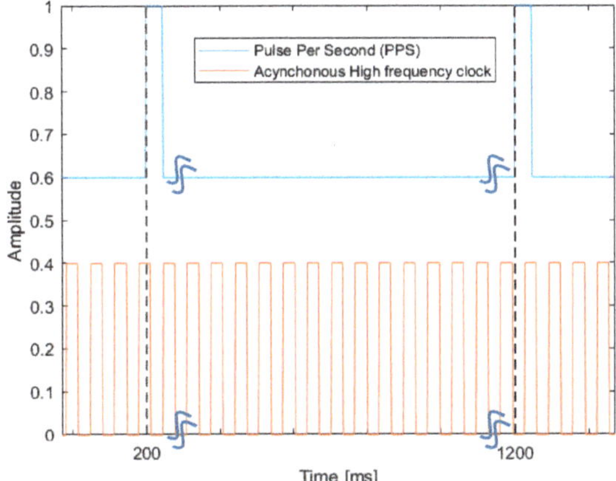


**Figure 7.** Example of an asynchronous high frequency clock.

### *4.2. PID Control*

As previously mentioned, to keep the internal TB phase locked with the PPS, a specific ISR takes the number of system ticks elapsed from the last PPS event and feeds a PID control algorithm. More in details, at each PPS event, the ISR snapshots the count value of the TBC, it computes the deviation, $e_j$, with respect to the values obtained at previous PPS event, and it feeds the PID. The output of the PID, divided by 50, will be the value used to configure the TBC rollover for the next second, as better explained in Section 4.3. In details, the new reload values $u_j$, at $j$-th iteration, can be computed as:

$$e_j = y_{ref} - y_j \tag{1}$$

$$u_{p,j} = K_p \cdot e_j;\ u_{i,j} = u_{i,j-1} + K_i \cdot e_j;\ u_{d,j} = K_d \cdot \left(e_j - e_{j-1}\right) \tag{2}$$

$$u_j = u_{p,j} + u_{i,j} + u_{d,j} \tag{3}$$

where $y_{ref}$ is the reference number of ticks, $y_j$ is the current value of counted ticks and $u_{p,j}$, $u_{i,j}$ and $u_{d,j}$ are, respectively, proportional, integrative and derivative components. Note that $u_{i,j}$ is computed recursively and $u_{d,j}$ is approximated as backward finite difference. Since the transfer function of the system is not known, the PID constants ($K_p$, $K_i$ and $K_d$) have been determined with the Ziegler-Nichols approach and empirically adjusted. The PID can act only one time per second, because reference time is only available one time per second (each PPS event). Moreover, a little value for $K_i$ constant has been chosen in order to obtain a low jitter in steady state conditions. This leads to a quite slow locking to absolute time, many iterations (between 60 and 80, depending on initial phase displacement) are needed to stably converge around the reference; thus, the anti-windup technique and the preload of integrator constant have been implemented to make the system work correctly within a few second after startup. In conclusion, the PID action keeps the TB of 50-pps locked in phase to the 1-PPS signal received by the GPS.

The accuracy of synchronization was experimentally evaluated: the time intervals between the active edges of the PPS signal and obtained time base periodicity have been measured with a 12-bit Lecroy MDA810 digital storage oscilloscope for about some hours. The measured distribution of the time delay is shown in Figure 8. Deviation, at steady state, exhibits almost a normal distribution with a mean systematic delay of 1.25 μs and maximum error of 500 ns. Thus, the system satisfactory reacts to internal clock drift and jitter obtaining good synchronization accuracy.

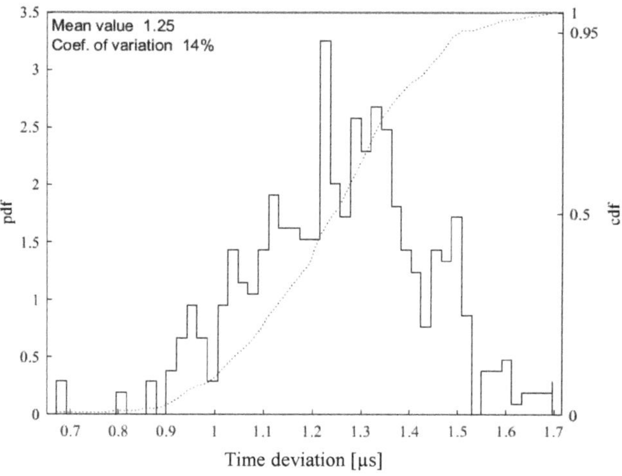


**Figure 8.** Measured synchronization accuracy.

### 4.3. Time-Base

The TB is a crucial part of the system as it has been used to trigger all the calculations. As previously described, the TB relies on an internal high-resolution timer, with a clock frequency as higher as possible to reach the best accuracy possible. High counter frequency implies that the counted number rapidly increases so that a 32-bit timer would be appropriate to implement the TBC. Unfortunately, a 32-bit timer running at system core clock is not a common feature on low cost MCUs; additionally, the adopted MCU has no 32-bit timers at 168 MHz but it has only 16-bit timers and it is not possible to perform a direct hardware cascade connection of two 16-bit timers (to obtain one single 32-bit timer from two 16-bit timers). To face these hardware limitations, a firmware technique to cascade two 16-bit counters for the TBC implementation has been implemented. When the first counter rollover occurs, an ISR manages the working of a second counter until the desired value is reached; particular attention has been paid to insure that the second counter stops before rollover. In the following, for sake of clarity, a numerical example is given for the explanation of the technique. Suppose that the system clock frequency is exactly equal to the nominal value, 168 MHz; in this case, to obtain the 20 ms periodicity for TB, it is needed to count up to 168,000,000/50 = 3,360,000 (reload value for TB), which is a number not representable with 16-bit. For this reason, the firmware lets the 16-bit counter to reach the rollup (so counting up to $2^{16}$) for a number of times equal to *floor* (3,360,000/$2^{16}$) = 51 and, on the

52-th iteration, the counter must be stopped before rollover; on the last iteration it must count 17,664, so reaching the desired value ($51 \times 2^{16} + 17664 = 3{,}360{,}000$). Another issue in the implementation of TB comes from the fact that the output of the PID, $u_j$, is the current tick number that corresponds to 1 s. This value has to be divided by 50 to obtain the TBC reload value to obtain a periodicity of 20 ms (50 fps). It is apparent that the result of the division is, in general, a decimal number while the TBC reload must be an integer number. A straightforward rounding of this value leads to a loss in resolution that could be relevant for the aimed application. Therefore, to mitigate this effect, the decimal part is properly considered: when, at the PPS event, a new TBC reload is calculated, in the next 50 TB events not always the same value is loaded. In fact, firmware reloads 50 potentially different TB values and the i-th reload value is calculated as:

$$reload\_value_i = \begin{cases} \left[\frac{u_j}{50}\right] & i = 0 \\ \left[\frac{u_j - \sum_{k=0}^{i-1} reload\_value_k}{50-i}\right] & i = 1,\ldots,49 \end{cases} \quad (4)$$

where square parenthesis means rounding to nearest integer. With this formula, the summation of the 50 TB values will always coincide with $u_j$.

### 4.4. AD Conversion

Two different ADC are used to acquire voltage and current signals and, to obtain simultaneous sampling, an internal timer has been selected as trigger source for both ADCs. The chosen sampling frequency was 6400 Hz as it corresponds to 128 samples per cycle at 50 Hz and the basic sampling time instants are triggered by a high-resolution timer running at system clock of 168 MHz. The desired sampling periodicity is obtained loading in the timer a proper counting value that nominally can be calculated dividing the rated system clock by the desired sampling frequency. However, the actual counting value has been experimentally evaluated. A square wave, with frequency equal to the desired sampling frequency has been generated. Its frequency has been measured with the frequency counter Agilent 53230A for some hours. The counting value has been chosen in such a way that the measured frequency of the square wave was as close as possible to 6.4 kHz; in particular, a sampling frequency of 6400.2500 Hz has been chosen, with a standard deviation of 100 μHz, which corresponds to about 0.02 μHz/Hz. The ADCs nominal resolution is 12-bit; nevertheless, for the purpose of the developed project, the effect of noise on the analog input or amplitude quantization can remarkably affect the overall accuracy of the results. Thus, ADC characteristics was enhanced through a technique based on over-sampling and averaging [40], thus obtaining improvement both in terms of amplitude resolution and noise rejection. Therefore, when a sampling time instant is triggered, the ADC is configured to acquire 16 samples at a burst sampling frequency much greater than the chosen sampling frequency (588 kHz). It is worthwhile noting that 16 samples are acquired in a really short time of about 27 μs, much lower than the equivalent sampling time interval. Then, the acquired samples are averaged to obtain the final sampling results. To this aim, two direct memory access channels are configured to simultaneously serve the two ADCs. They copy the samples in dedicated queues and an ISR averages the 16 samples of voltage and current, respectively, using the integrated DSP unit. The averaged results are pushed in two different queues, one for voltage and one for current. A final issue is related to the time collocation of the averaged values: they are the result of the averaging of 16 samples over 27 μs. They are considered at the center of the averaging interval and so a systematic phase delay is introduced; however, this phase delay can be compensated with a correction of the measured synchrophasor phase value returned by the estimation algorithm.

### 4.5. Resampling

As previously described, to correctly measure a synchrophasor, the input signals must be synchronously sampled with the UTC. In the considered prototype, the sampling time instants are asynchronous with UTC, so they cannot be used directly for synchrophasor calculation for different reasons: (1) the number of samples is different from nominal value (128 samples) in each 20 ms frame, (2) the actual number of samples that corresponds to 20 ms is not an integer number of samples and (3) the first sample of each frame is not aligned with an event of TB. Therefore, the desynchronized acquired signal is resampled with linear interpolation: we can calculate an approximation of the synchronized sample by interpolating the two acquired samples that are closer to the desired time position. Obviously, a more complex interpolation technique can improve the overall accuracy of the instrument. The first step for the resampling is to put in relation the desynchronized and synchronized time instants. The synchronized time instants $t_s(i)$ can be easily obtained starting from TB events, with a uniform spacing of $T_c$ = 20 ms/128 (see Figure 9). The desynchronized time instants, $t_d(j)$, are spaced of sampling period $T_s[k]$ estimated in the $k$-th frame. The data needed for estimating $T_s[k]$ are collected by the TB ISR in CI and they are placed in a dedicated queue. In particular, CI includes:

- $N[k]$, the number of samples acquired in the $k$-th frame;
- $L_{tick}[k]$, the value of the system clock counts (systick) read when the last event of the TB occurs;
- $PPS_{tick}$, the systick read when the last PPS event occurs.

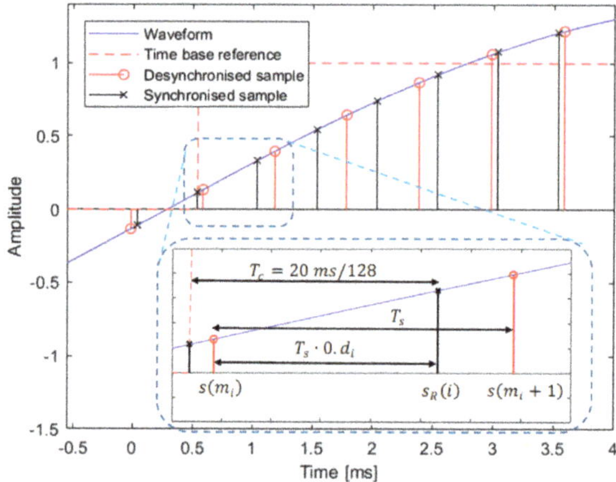


**Figure 9.** Resampling of voltage signal; actual samples are interpolated in desired instants of time.

So $T_s[k]$ can be estimated as:

$$T_s[k] = \frac{L_{tick}[k] - L_{tick}[k-1]}{N[k]} = \frac{1}{f_s[k]} \quad (5)$$

where $f_s[k]$ is the sampling frequency estimated in the $k$-th frame. With this information, it is possible to find out what are the nearest acquired samples, by calculating where the synchronized samples should be "virtually" located in the acquisition buffer. These "virtual locations" of the i-th synchronized sample, $n_i$, come out by dividing the synchronized time instants, $t_s(i) = i{\cdot}20$ ms, by the measured sampling period, $T_s[k]$:

$$n_i = \frac{t_s(i)}{T_s[k]} = i{\cdot}\frac{20\text{ ms}}{T_s[k]} = m_i.d_i \quad (6)$$

where $n_i$ are not-integer indexes, with an integer part, $m_i$, and a decimal part, $d_i$. Thus, the i-th synchronized sample should have been acquired between the actual acquired samples, that are at indexes $m_i$ and $m_i + 1$ of the buffer, with a normalized distance from the former of $0.d_i$ and with a distance from the latter of $1 - 0.d_i$ (see Figure 4). Therefore, an estimation of the $i$-th synchronized sample, $s_R(i)$, can be calculated as average of two nearest samples weighted according relative distances thus:

$$s_R(i) = s(m_i) \cdot (1 - 0.d_i) + s(m_i + 1) \cdot 0.d_i \tag{7}$$

where $s(m_i)$ and $s(m_i + 1)$ are two subsequent acquired samples, closest to the $i$-th synchronized time instants (see Figure 9). With this approach, 512 synchronized samples are calculated so that, the first half is taken before, and the other half is taken after each synchronized TB event.

### 4.6. Synchrophasor Calculation

To estimate the synchrophasor, different techniques were proposed [13–19] in scientific literature. Most of them are eligible to be implemented in the proposed prototype; for this paper, the technique based on interpolation in the frequency domain is chosen [14]. The algorithm starts with calculating Discrete Fourier Transform (DFT) of resampled values. These samples are synchronous with absolute time but, in general, asynchronous with respect to the power system frequency. Thus, none of the calculated spectral components exactly matches with the fundamental components of the acquired signals. Nevertheless, its actual amplitude, phase angle and frequency can be evaluated by interpolating the obtained spectral components. This procedure can be adopted on the assumption of negligibility of the spectral leakage effects due to other sinusoidal components and due to spectral replica at negative frequency. This assumption can be made adopting an opportune window like Hanning window that has good performance, relatively to attenuation of spectral leakage [14]. For the case at hand, the effect of negative replica results to be prevalent and iterative estimation procedure was adopted to make this effect negligible.

### 4.7. Execution Times

All the execution times of the different routines of the implemented firmware have been measured. In particular, the maximum reporting rate has been considered, that is 50 Hz, and an observation interval of four cycles of nominal power frequency (50 Hz) has been used. Each execution time has been measured in this way: a digital pin of the microcontroller has been set to high value just before the execution of the specific routine and again set to low value just after the routine execution; the duration of the obtained pulse was measured with the 12-bit Lecroy MDA810 digital storage oscilloscope. The results are as follows: (1) PID routine takes 1.6 μs every 1 s, (2) TB routine takes 2.5 μs every 20 ms (50 times a second), (3) oversampling and averaging take 3 μs every 156 μs (6400 times a second), (4) extraction from queue routine takes 2.2 μs every 20 ms (50 times a second), (5) time domain interpolation takes 6.8 ms every 20 ms (50 times a second), (6) IpDFT takes 560 μs every 20 ms (50 times a second). Summarizing, the total execution time is 7.8 ms for the two channels, i.e., quite lower than 20 ms, which is the time interval between two reporting instants.

## 5. Experimental Results

The realized PMU prototype has been tested with a high performance PMU calibrator, the Fluke 6135A/PMUCAL.

A metrological characterization of the calibrator has been performed, by some of the authors, in [38]. It is able to give reliable results down to values of 0.012% for TVE, 0.6 mHz for Frequency Error (FE) and 0.07 Hz/s for Rate of change of Frequency Error (RFE), in the various test conditions of [2,3].

The presented prototype has been tested in several testing conditions, both for class P as well as for class M PMUs, as prescribed in [2,3], for rated voltage of 230 V and rated current of 10 A. The chosen testing conditions are:

1. Sinewaves with off-nominal frequency deviations within ±2 Hz (class P) and ±5 Hz (class M);

2. Sinewaves with off-nominal frequency deviations within ±2 Hz (class P) and ±5 Hz (class M) affected by one single harmonic component, from 2nd to 50th, of amplitude equal to 1% (class P) and 10% (class M) of the fundamental;
3. Amplitude Modulated (AM) sinewaves affected by a 2 Hz (class P) and 5 Hz (class M) modulating tone of amplitude equal to 10% of the fundamental;
4. Phase Modulated (PM) sinewaves affected by a 2 Hz (class P) and 5 Hz (class M) modulating tone of amplitude equal to 0.1 rad;
5. 50 Hz sinewaves corrupted by a single out-of-band inter-harmonic (only class M) of amplitude equal to 10% of the fundamental at 24.9 Hz;
6. Chirp waveforms with the fundamental frequency increasing linearly from about 48 Hz to 52 Hz (class P) and from 45 Hz to 55 Hz (class M), and vice versa, at a rate of ±1 Hz/s.

Results of the experimental tests are summarized in Table 1 (class P) and Table 2 (class M). They were obtained using an observation interval of four nominal 50 Hz cycles and a reporting rate of 50 fps. For sake of brevity, only results relative to voltage channel are presented; similar values have been obtained also for current channel.

**Table 1.** Maximum measured (Meas.) TVE, FE and RFE, and the corresponding limit values, in various testing conditions reported in IEEE Standards for Class P PMUs. The reported results refer to observation intervals of four nominal cycles and reporting rate of 50 fps.

| Test Type | TVE max [%] | | FE max [mHz] | | RFE max [Hz/s] | |
|---|---|---|---|---|---|---|
| | Limit | Meas. | Limit | Meas. | Limit | Meas. |
| **Frequency offset (±2 Hz)** | 1 | 0.11 | 5 | 1.0 | 0.4 | 0.13 |
| **Frequency offset (±2 Hz) +1% 2nd harmonic** | 1 | 0.19 | 5 | 4.7 | 0.4 | 0.19 |
| **Frequency offset (±2 Hz) +1% 3rd harmonic** | 1 | 0.15 | 5 | 1.1 | 0.4 | 0.15 |
| **Frequency ramp (±2 Hz at 1 Hz/s)** | 1 | 0.35 | 10 | 7.7 | 0.4 | 0.19 |
| **AM (10% at 2 Hz)** | 3 | 1.8 | 60 | 27.0 | 2.3 | 2.1 |
| **PM (0.1 rad at 2 Hz)** | 3 | 1.4 | 60 | 45.0 | 2.3 | 1.5 |

**Table 2.** Maximum measured (Meas.) TVE, FE and RFE, and the corresponding limit values, in various testing conditions reported in IEEE Standards for Class M PMUs. The reported results refer to observation intervals of four nominal cycles and reporting rate of 50 fps.

| Test Type | TVE max [%] | | FE max [mHz] | | RFE max [Hz/s] | |
|---|---|---|---|---|---|---|
| | Limit | Meas. | Limit | Meas. | Limit | Meas. |
| **Frequency offset (±5 Hz)** | 1 | 0.21 | 5 | 1.0 | 0.1 | 0.14 |
| **Frequency offset (±5 Hz) +1% 2nd harmonic** | 1 | 0.22 | 25 | 9.5 | - | 0.45 |
| **Frequency offset (±5 Hz) +1% 3rd harmonic** | 1 | 0.21 | 25 | 1.2 | - | 0.16 |
| **Frequency ramp (±5 Hz at 1 Hz/s)** | 1 | 0.44 | 10 | 8.7 | 0.2 | 0.21 |
| **AM (10% at 5 Hz)** | 3 | 2.3 | 300 | 153.0 | 14 | 12 |
| **PM (0.1 rad at 5 Hz)** | 3 | 2.5 | 300 | 253.0 | 14 | 14 |
| **10% out-of-band inter-harmonic @ ≈ 25 Hz** | 1.3 | 2.7 | 10 | 854.0 | - | 45 |

The IpDFT algorithm generally shows good performance in steady-state conditions, in particular in presence of harmonics and inter-harmonics. Nevertheless, the performance is worse in dynamic conditions (with frequency ramp and especially with amplitude and phase modulations). These results are essentially due to the static phasor model at the base of the IpDFT and become worse with the increase of the time observation window.

However, it can be seen that the TVE, FE and RFE are below the standard limits for practically all the testing conditions, except for the out-of-band interharmonic, where both TVE and FE are over the limit, and for the frequency ramp for class M, where the RFE is slightly worse (0.21 Hz/s) than the limit (0.20 Hz/s).

These results are particularly relevant especially if compared with the results shown in [38], where the experimental characterization of the IpDFT algorithm is performed using a high-performance measuring hardware, constituted by a PXI controller, 16-bit data acquisition system, high accuracy class voltage and current transducers and a synchronization board acting as GPSDO.

The performance here shown is, as expected, worse, due to the intrinsic poor performance of the hardware here used; however, some situations, where errors are near (or over) the limit, are highlighted also in [38], thus depending essentially on the used estimation algorithm.

## 6. Conclusions

In this paper, a design approach for the realization of a very low cost PMU is presented. The core of the instrument is an ARM microcontroller with integrated ADC and Ethernet controller. External devices are used to realize the voltage and current sensing stage, a simple GPS receiver is used to receive the PPS and the timestamp and, in particular, no GPS disciplined oscillator has been used.

Since a key feature of the PMU is the accurate synchronization with UTC, in order to meet this requirement with a sufficient accuracy, and without specific synchronization hardware, an efficient signal processing synchronization technique has been used, which is able to lock the sampling frequency to the UTC. To verify the robustness of the proposed design approach, a very common phasor estimation algorithm, the IpDFT, has been implemented onboard the microcontroller. Its features are the ease to implement and the quite low computational complexity.

The cost of the prototype is very low, about 110 €, obtained in this way: (1) the transducer cost is about 90 €, (2) the cost of the microcontroller development board is about 10 €, (3) the cost of the GPS receiver and the antenna is lower than 10 € and (4) the electronic components of the conditioning circuit is lower than 1 €.

The realized prototype has been tested with a high performance PMU calibrator, the Fluke 6135A/PMUCAL, in several testing conditions reported in [2,3]. The maximum values for TVE, FE and RFE are below the standard limits practically for every testing condition, except for interharmonic, and for RFE in the frequency ramp test for Class M PMUs. However, comparing the obtained results with those obtained from an IpDFT implementation on a high performance measuring hardware [38], nearly the same issues have been found.

**Author Contributions:** Conceptualization, M.L.; methodology, C.L. and D.G.; software, A.D.F.; validation, M.L. and A.D.F.; formal analysis, C.L. and D.G.; investigation, M.L., C.L., D.G. and A.D.F.; resources, M.L. and C.L.; data curation, A.D.F.; writing—original draft preparation, M.L. and A.D.F.; writing—review and editing, C.L. and D.G.; supervision, C.L.; funding acquisition, M.L. and C.L.

**Funding:** The work presented in this paper was funded by European Metrology Programme for Innovation and Research (EMPIR), 17IND06 Future Grid II project, which is jointly funded by the EMPIR participating countries within EURopean Association of national METrology institutes (EURAMET) and the European Union.

**Conflicts of Interest:** The authors declare no conflict of interest.

## References

1. Bose, A. Smart transmission grid applications and their supporting infrastructure. *IEEE Trans. Smart Grid* **2010**, *1*, 11–19. [CrossRef]

2. IEEE Standard for Synchrophasor Measurements for Power Systems. In *IEEE Std C37.118.1-2011 (Revision of IEEE Std C37.118-2005)*; IEEE: Piscataway, NJ, USA, 2011; pp. 1–61. [CrossRef]
3. IEEE Standard for Synchrophasor Measurements for Power Systems—Amendment 1: Modification of Selected Performance Requirements. In *IEEE Std C37.118.1a-2014 (Amendment to IEEE Std C37.118.1-2011)*; IEEE: Piscataway, NJ, USA, 2014; pp. 1–25. [CrossRef]
4. De La Ree, J.; Centeno, V.; Thorp, J.S.; Phadke, A.G. Synchronized Phasor Measurement Applications in Power Systems. *IEEE Trans. Smart Grid* **2010**, *1*, 20–27. [CrossRef]
5. Silverstein, A. Synchrophasors & the Grid. In Proceedings of the Electricity Advisory Committee Meeting, Arlington, VA, USA, 13 September 2017.
6. U.S. Department of Energy. *Advancement of Synchrophasor Technology in Projects Funded by the American Recovery and Reinvestment Act of 2009*; U.S. Department of Energy: Washington, DC, USA, 2016.
7. U.S. Department of Energy. *Office of Electricity Delivery and Energy Reliability, Factors Affecting PMU Installation Costs*; U.S. Department of Energy: Washington, DC, USA, 2014.
8. Lelic, D. PMU Cost & Benefits Study. In Proceedings of the Center for Advanced Power Engineering Research Meeting. Available online: http://caper-usa.com/wp-content/uploads/2017/08/Session-IV-PMU-Cost-Benefits-Studies-DLelic_8-8-17.pdf (accessed on 10 July 2019).
9. Baldwin, T.L.; Mili, L.; Boisen, M.B.; Adapa, R. Power system observability with minimal phasor measurement placement. *IEEE Trans. Power Syst.* **1993**, *8*, 707–715. [CrossRef]
10. Nuqui, R.F.; Phadke, A.G. Phasor measurement unit placement techniques for complete and incomplete observability. *IEEE Trans. Power Deliv.* **2005**, *20*, 2381–2388. [CrossRef]
11. Zhong, Z.; Xu, C.; Billian, B.J.; Zhang, L.; Tsai, S.-J.S.; Conners, R.W.; Centeno, V.A.; Phadke, A.G.; Liu, Y. Power system frequency monitoring network (FNET) implementation. *IEEE Trans. Power Syst.* **2005**, *20*, 1914–1921. [CrossRef]
12. Phadke, A.G.; Kasztenny, B. Synchronized Phasor and Frequency Measurement Under Transient Conditions. *IEEE Trans. Power Deliv.* **2009**, *24*, 89–95. [CrossRef]
13. De la O Serna, J.A.; Platas, M.A. Maximally flat differentiators through WLS Taylor decomposition. *Elsevier Digit. Signal Process.* **2011**, *21*, 183–194. [CrossRef]
14. Belega, D.; Petri, D. Accuracy Analysis of the Multicycle Synchrophasor Estimator Provided by the Interpolated DFT Algorithm. *IEEE Trans. Instrum. Meas.* **2013**, *62*, 942–953. [CrossRef]
15. Petri, D.; Fontanelli, D.; Macii, D. A Frequency-Domain Algorithm for Dynamic Synchrophasor and Frequency Estimation. *IEEE Trans. Instrum. Meas.* **2014**, *63*, 2330–2340. [CrossRef]
16. Bertocco, M.; Frigo, G.; Narduzzi, C.; Muscas, C.; Pegoraro, P.A. Compressive Sensing of a Taylor-Fourier Multifrequency Model for Synchrophasor Estimation. *IEEE Trans. Instrum. Meas.* **2015**, *64*, 3274–3283. [CrossRef]
17. Cuccaro, P.; Gallo, D.; Landi, C.; Luiso, M.; Romano, G. Recursive phasor estimation algorithm for synchrophasor measurement. In Proceedings of the 2015 IEEE International Workshop on Applied Measurements for Power Systems (AMPS), Aachen, Germany, 23–25 September 2015; pp. 90–95.
18. Cuccaro, P.; Gallo, D.; Landi, C.; Luiso, M.; Romano, G. Phase-based estimation of synchrophasors. In Proceedings of the 2016 IEEE International Workshop on Applied Measurements for Power Systems (AMPS), Aachen, Germany, 28–30 September 2016; pp. 1–6.
19. Tosato, P.; Macii, D.; Luiso, M.; Brunelli, D.; Gallo, D.; Landi, C. A Tuned Lightweight Estimation Algorithm for Low-Cost Phasor Measurement Units. *IEEE Trans. Instrum. Meas.* **2018**, *67*, 1047–1057. [CrossRef]
20. Jiang, J.A.; Yang, J.Z.; Lin, Y.H.; Liu, C.W.; Ma, J.C. An adaptive PMU based fault detection/location technique for transmission lines. I. Theory and algorithms. *IEEE Trans. Power Deliv.* **2000**, *15*, 486–493. [CrossRef]
21. Tate, J.; Overbye, T.J. Line Outage Detection Using Phasor Angle Measurements. *IEEE Trans. Power Syst.* **2008**, *23*, 1644–1652. [CrossRef]
22. Von Meier, A.; Culler, D.; McEachern, A.; Arghandeh, R. Microsynchrophasors for distribution system. In Proceedings of the 5th IEEE PES Innovative Smart Grid Technologies Conference (ISGT), Washington, DC, USA, 19–22 February 2014; pp. 1–5.
23. Romano, P.; Paolone, M. Enhanced interpolated-DFT for synchrophasor estimation in FPGAs: Theory, implementation, and validation of a PMU prototype. *IEEE Trans. Instrum. Meas.* **2014**, *63*, 2824–2836. [CrossRef]
24. Das, H.P.; Pradhan, A.K. Development of a micro-phasor measurement unit for distribution system applications. In Proceedings of the National Power Systems Conference (NPSC), Bhubaneswar, India, 19–21 December 2016; pp. 1–5.

25. Laverty, D.M.; Best, R.J.; Brogan, P.; al Khatib, I.; Vanfretti, L.; Morrow, D.J. The OpenPMU Platform for Open-Source Phasor Measurements. *IEEE Trans. Instrum. Meas.* **2013**, *62*, 701–709. [CrossRef]
26. Zhao, X.; Laverty, D.M.; McKernan, A.; Morrow, D.J.; McLaughlin, K.; Sezer, S. GPS-Disciplined Analog-to-Digital Converter for Phasor Measurement Applications. *IEEE Trans. Instrum. Meas.* **2017**, *66*, 2349–2357. [CrossRef]
27. Yao, Y.; Zhan, L.; Liu, Y.; Till, M.J.; Zhao, J.; Wu, L.; Teng, Z. A Novel Method for Phasor Measurement Unit Sampling Time Error Compensation. *IEEE Trans. Smart Grid* **2018**, *9*, 1063–1072.
28. Cataliotti, A.; di Cara, D.; Emanuel, A.E.; Nuccio, S. Improvement of Hall Effect Current Transducer Metrological Performances in the Presence of Harmonic Distortion. *IEEE Trans. Instrum. Meas.* **2010**, *59*, 1091–1097. [CrossRef]
29. Faifer, M.; Laurano, C.; Ottoboni, R.; Toscani, S.; Zanoni, M. Characterization of Voltage Instrument Transformers Under Nonsinusoidal Conditions Based on the Best Linear Approximation. *IEEE Trans. Instrum. Meas.* **2018**, *67*, 2392–2400. [CrossRef]
30. Crotti, G.; Delle Femine, A.; Gallo, D.; Giordano, D.; Landi, C.; Letizia, P.S.; Luiso, M. Calibration of Current Transformers in distorted conditions. In *Journal of Physics: Conference Series*; IOP Publishing: Bristol, UK, 2018. [CrossRef]
31. Del Prete, S.; Delle Femine, A.; Gallo, D.; Landi, C.; Luiso, M. Implementation of a distributed Stand Alone Merging Unit. In *Journal of Physics: Conference Series*; IOP Publishing: Bristol, UK, 2018. [CrossRef]
32. Crotti, G.; Delle Femine, A.; Gallo, D.; Giordano, D.; Landi, C.; Luiso, M. Measurement of the Absolute Phase Error of Digitizers. *IEEE Trans. Instrum. Meas.* **2019**, *68*, 1724–1731. [CrossRef]
33. Crotti, G.; Giordano, D.; Delle Femine, A.; Gallo, D.; Landi, C.; Luiso, M. A Testbed for Static and Dynamic Characterization of DC Voltage and Current Transducers. In Proceedings of the 9th IEEE International Workshop on Applied Measurements for Power Systems (AMPS), Bologna, Italy, 26–28 September 2018; pp. 1–6. [CrossRef]
34. Collin, A.J.; Delle Femine, A.; Gallo, D.; Langella, R.; Luiso, M. Compensation of Current Transformers' Non-Linearities by Means of Frequency Coupling Matrices. In Proceedings of the 9th IEEE International Workshop on Applied Measurements for Power Systems (AMPS), Bologna, Italy, 26–28 September 2018; pp. 1–6. [CrossRef]
35. Delle Femine, A.; Gallo, D.; Giordano, D.; Landi, C.; Luiso, M.; Signorino, D. Synchronized Measurement System for Railway Application. In *Journal of Physics: Conference Series*; IOP Publishing: Bristol, UK, 2018. [CrossRef]
36. Cataliotti, A.; Cosentino, V.; Crotti, G.; Femine, A.D.; Di Cara, D.; Gallo, D.; Giordano, D.; Landi, C.; Luiso, M.; Modarres, M.; et al. Compensation of Nonlinearity of Voltage and Current Instrument Transformers. *IEEE Trans. Instrum. Meas.* **2019**, *68*, 1322–1332. [CrossRef]
37. Crotti, G.; Femine, A.D.; Gallo, D.; Giordano, D.; Landi, C.; Luiso, M.; Mariscotti, A.; Roccato, P.E. Pantograph-to-OHL Arc: Conducted Effects in DC Railway Supply System. In Proceedings of the 9th IEEE International Workshop on Applied Measurements for Power Systems (AMPS), Bologna, Italy, 26–28 September 2018; pp. 1–6. [CrossRef]
38. Luiso, M.; Macii, D.; Tosato, P.; Brunelli, D.; Gallo, D.; Landi, C. A Low-Voltage Measurement Testbed for Metrological Characterization of Algorithms for Phasor Measurement Units. *IEEE Trans. Instrum. Meas.* **2018**, *67*, 2420–2433. [CrossRef]
39. IEEE Standard for Synchrophasor Data Transfer for Power Systems. In *IEEE Std C37.118.2-2011 (Revision of IEEE Std C37.118-2005)*; IEEE: Piscataway, NJ, USA, 2011; pp. 1–53. [CrossRef]
40. AN4073. ST Application Note How to Improve ADC Accuracy When Using STM32F2xx and STM32F4xx Microcontrollers DocID022945 Rev 5 July 2013. Available online: https://www.st.com/content/ccc/resource/technical/document/application_note/a0/71/3e/e4/8f/b6/40/e6/DM00050879.pdf/files/DM00050879.pdf/jcr:content/translations/en.DM00050879.pdf (accessed on 10 July 2019).

*Article*

# An Optimized HT-Based Method for the Analysis of Inter-Area Oscillations on Electrical Systems

**Francesco Bonavolontà [1,†], Luigi Pio Di Noia [1,†], Davide Lauria [2,†] and Annalisa Liccardo [1,*,†] and Salvatore Tessitore [1,†]**

1 Department of Electrical Engineering and Information Technology, University of Naples Federico II, 80125 Napoli, Italy
2 Department of Industrial Engineering, University of Naples Federico II, 80125 Napoli, Italy
* Correspondence: annalisa.liccardo@unina.it; Tel.: +39-081-768-3912
† These authors contributed equally to this work.

Received: 6 July 2019; Accepted: 27 July 2019; Published: 31 July 2019

**Abstract:** The paper deals with a novel method to measure inter-area oscillations, i.e., electromechanical oscillations involving groups of generators geographically distant from one another and ranging within the frequency interval from 0.1 Hz up to 1 Hz. The estimation of the parameters characterizing inter-area oscillations is a crucial objective in order to take the necessary actions to avoid the instability of the transmission electrical system. The proposed approach is a signal-based method, which uses samples of electrical signals acquired by the phasor measurement unit (PMU) and processes them to extract the individual oscillations and, for each of them, determine their characteristic parameters such as frequency and damping. The method is based on Hilbert transformations, but it is optimized through further algorithms aiming at (i) improving the ability to separate different oscillatory components, even at frequencies very close to each other, (ii) enhancing the accuracy associated with the damping estimates of each oscillation, and (iii) increasing the robustness to the noise affecting the acquired signal. Results obtained in the presence of signals involving the composition of two oscillations, whose damping and frequency have been varied, are presented. Tests were conducted with signals either synthesized in simulated experiment or generated and acquired with actual laboratory instrumentation.

**Keywords:** damping estimation; Hilbert transform; oscillations reshaping; nonlinear least square approximation

## 1. Introduction

Inter-area oscillations are phenomena that affect electricity systems interconnecting areas geographically distant from each other, such as different nations. Various phenomena can cause the inter-area oscillation; a rapid change of load, a line faults, the control and regulation necessary to guarantee the stability of large power system area, the connection between large power system with weak lines. Practically, all these phenomena determine a transient in the transferred power which bring the system into a new balance operating condition [1].

The balance point will be reached after a redefinition of the power contributions of the various generators connected to the system [2].

The dynamic modes characterizing the motion from a balance condition to another belong to the Low Frequency Oscillations (LFOs) category. In particular, each $i$-th mode of evolution of the electrical system is typically representable according to the following expression:

$$y_i(t) = A_i e^{-d_i t} \sin(2\pi f_i t + \varphi_i), \quad (1)$$

where $A_i$ is the amplitude, $d_i$ is the damping coefficient, $f_i$ is the frequency and $\varphi_i$ is the starting phase.

Usually, the frequency range is from 0.1 Hz up to 2 Hz and the oscillations are divided into two categories:

- inter-area oscillations from 0.1 to 1 Hz;
- local oscillations from 1 up to 2 Hz.

As regards the damping value, a generic classification adopted in the scientific literature is the following [3–5]:

- damped $d_i \in [0.05, +\infty[$;
- weakly damped $d_i \in ]0, 0.05[$;
- divergent $d_i \in ]-\infty, 0[$.

The accurate estimate of the damping coefficient of the oscillations is a fundamental issue. In fact, particularly dangerous for the stability of the electrical system are the oscillations that exhibit a weakly positive or negative damping coefficient. In fact, if the oscillations are not attenuated for a long time or they are amplified and can reach dangerous amplitudes, they can cause the intervention of protections or, in the worst cases, could jeopardize network operation.

The estimation of damping coefficient is crucial to take the adequate countermeasures in order to assure the network stability. On the other hand, the inter-area oscillations are often characterized by weak damping, and the estimation of the damping sign is not a straightforward process.

In the paper, a novel method based on an optimized combination between the Hilbert transform (HT) and non-linear least square (NLS) fitting is proposed. The method aims at separating the dynamic modes that are combined in the acquired signal and estimating frequency and damping coefficient of the individual oscillations.

The paper is organized as follows. In Section 2 the main methods described in the literature for the analysis of inter-area oscillations are discussed. In Section 3 theoretical fundamentals about HT are given. In Section 4 the proposed approach is presented. The results obtained from both numerical and experimental assessment are reported in Section 5. Section 6 deals with the digital signal processing method to analyze oscillations characterized by very close frequency values. Concluding remarks are given in Section 7.

## 2. Approaches Available In Literature

Recognizing as soon as possible the divergent or weakly damped oscillations is a fundamental task to be able of employing appropriate countermeasures and consequently guarantee the stability of the electrical systems. The estimation algorithm has to exhibit, as main characteristics, a low computational burden, high robustness to noise, high accuracy in parameter estimation and no need of a priori information about the signal of interest.

Generally, the methodologies reported in scientific literature follow two different approaches [6]:

- Model-based;
- Signal-based.

As reported in [6–8] the model-based approach aims at estimating the parameters of synchronous generators in order to derive an overall model of the electrical system. The possibility of retrieving the system model is appealing, but these methods are characterized by high computational burden. The order of the equivalent model, in fact, can be very high, since it is strictly dependent on the extension of the electrical system [9]. Moreover, the obtained equivalent model is valid for a limited time interval because of the constant evolution of the network topology.

Although many methods have been proposed aiming at power system reduction for dynamic analysis with consequent lower computational burden, the results obtained do not seem to be sufficiently accurate.

An interesting approach is represented by a hybrid method based on the Kalman Filter, reported in [10,11], that combines the model-based and signal-based methodologies. This method processes signals acquired through the phasor measurement units (PMUs) to obtain the overall model of the electrical system. As known, the Kalman filter is a valuable approach for real-time signal analysis, thanks to its adaptive behavior. However, also this method exhibits a significant computational burden, because of the matrices dimensions, dependent on the complexity of the model. For this reason, this approach is usually adopted to estimate damping coefficient and frequency of the dominant mode, with consequent loss of information about eventual other oscillatory modes. Moreover, the proper initial choice of the measurement noise covariance and process noise covariance is the most sensitive issue characterizing the Kalman filter.

Owing to the analytical difficulties involved in Model-based approach, the scientific literature is increasingly oriented towards the study and development of a Signal-based method for the analysis of LFOs.

The typical structure of such algorithms is characterized by two steps: decomposition in mono-component signals, and estimation of signal parameters.

Several solutions available in the literature match with this type of approach: variational mode decomposition (VMD) and Teager–Kaiser in [12]; discrete Fourier transform (DFT) and curve-fitting in frequency domain in [13]; singular value decomposition (SVD) and matrix pencil method in [14]; empirical mode decomposition (EMD) and Hilbert spectrum analysis in [15,16], as an example. Each of the mentioned algorithms is characterized by different advantages and drawbacks. Generally, the most critical issue is related to the non-stationary and non-linear nature of the electrical system [17].

The most analyzed approach, in the literature, is undoubtedly the Prony method [3]. The Prony algorithm is a parametric method of direct signal decomposition. It assumes that the acquired signal can be approximated by a sum of damped sinusoids. This method does not need a priori information, since a precise model has been chosen to fit the signal; moreover it only needs to analyze a period and a half of the oscillations in order to estimate its damping and frequency. However, as is well known in the literature, the algorithm is very sensitive to the noise, so that in some cases additional mono-component signals are not detected. The direct consequence is a limited accuracy in the estimation of the parameters of interest. In order to solve this problem, pre-filtering techniques are used for noise cancellation, such as digital filters and EMD, causing an increase of the computational burden of the algorithm.

In [18] the authors propose an improved Prony method where the method is coupled with a low order model of the signal and the parameters are evaluated with a least square method. Instead, the paper [19] introduces a covariance matrix in order to improve the estimation of low frequencies with Prony method.

Another interesting approach involves the use of multivariate autoregressive model which helps in the identification of low frequencies signals. In [20] the identification of inter-area oscillations is carried out through the use of Yule–Walker estimator; the results highlight the good performance of the algorithm also when the measurement noise and the lost of packets affects the signal.

Authors have finally focused their attention on Hilbert-based approaches as in [21,22]. These methods are flexible, do not require information about the signal and are suitable for representing non-linear and non-stationary systems. The critical issue is the process for separating the mono-component oscillations; even a small edge effect causes a considerable lack of accuracy in the damping estimate.

## 3. Ht Fundamentals

The HT of a real signal $x(t)$ is defined as the following convolution integral:

$$H[x(t)] = \frac{1}{\pi} P.V. \int_{+\infty}^{-\infty} \frac{x(\tau)}{t-\tau} d\tau, \tag{2}$$

where it is necessary to consider the Cauchy Principal Value ($P.V.$) of the integral, resulting from the eventual singularity of the integrand function.

The HT has the fundamental property of shifting the phase of the real signal of a positive angle equal to $\frac{\pi}{2}$, without the need of changing the representation domain (both HT input and output belong to the time domain). This characteristic is useful to obtain the analytic signal of a real signal $x(t)$.

$$z(t) = x(t) + jx_H(t) = x(t) + jH[x(t)] \tag{3}$$

The signal can be written according to the polar coordinates, so it can be seen as the product between two terms: the first one takes into account the amplitude modulation and the second one includes the phase modulation. If the signal exhibits slow amplitude variations with respect to the phase variations, i.e., if the amplitude modulation is characterized by frequencies much lower then those characterizing the phase modulation [23], the analytic signal in polar coordinates can be considered as a rotating phasor:

$$z(t) = A(t)\, e^{j\theta(t)}. \tag{4}$$

The time-varying amplitude is the envelope of the real signal and phase can be estimated according to the equations:

$$A(t) = \sqrt{x(t)^2 + H[x(t)]^2} \tag{5}$$

$$\theta(t) = \arctan \frac{H[x(t)]}{x(t)}. \tag{6}$$

From the expression of the instantaneous phase it is possible to determine the instantaneous frequency, defined as:

$$f(t) = \frac{1}{2\pi}\frac{d\theta_u(t)}{dt}, \qquad \theta_u(t) = \theta(t) + \Lambda(t), \tag{7}$$

where $\Lambda(t)$ is a function whose values are integer multiples of $\pi$, required to assure the continuity of the phase $\theta_u(t)$.

If $x(t)$ is a damped oscillation, the HT is a suitable tool for estimating the oscillation parameters. Equation (7), in fact, provides the oscillation frequency; the instantaneous amplitude, instead, provides the damping factor. In fact, the instantaneous amplitude can be written as:

$$A(t) = A_1 e^{-\sigma t} \tag{8}$$

where $A_1$ is the starting amplitude and $\sigma$ is the damping coefficient. By performing the natural logarithm:

$$\ln[A(t)] = \ln(A_1) - \sigma t \tag{9}$$

The damping coefficient, thus, can be evaluated performing the time derivative of the terms of Equation (9):

$$\sigma = \frac{d}{dt}(\sigma t) = -\frac{d}{dt}[\ln(A(t))]. \tag{10}$$

Since HT is a linear operation, the following equation holds:

$$H[kx(t)] = kH[x(t)] \tag{11}$$

where $k$ is a generic constant.

By keeping this in mind and by recalling the property of shifting the phase, if $x(t)$ is a cosinusoidal function, according to Equation (11), its HT is:

$$H[A\cos(\omega t + \varphi)] = A\sin(\omega t + \varphi), \tag{12}$$

where $A$ is the constant amplitude, $\omega$ stands for the angular frequency, $\varphi$ represents the phase and $\sin(\omega t + \varphi)$ is the HT of the cosine. This relationship will be exploited in the successive section.

Lastly, if $x(t)$ is the product of two functions, namely $f$ and $g$, the HT of $x(t)$ could be evaluated as the multiplication between $f(t)$ and the HT of $g(t)$, according to:

$$H[f(t) \cdot g(t)] = f(t) \cdot H[g(t)]. \tag{13}$$

This equivalence is maintained if $f$ and $g$ satisfy the Bedrosian's conditions [16,21], i.e., if their spectra are separated and $f$ is characterized by low-frequency components with respect to $g$.

## 4. Proposed Approach

The proposed method is event-based, i.e., the algorithm has to run when an oscillation is detected, in order to estimate its parameters; in normal operating conditions, the results of the parameters estimation would be unmeaningful.

The proposed approach consists of three main steps:

- Separation of individual oscillations: the acquired signal is processed in order to recognize the individual inter-area oscillations;
- Oscillations reshaping: the estimation of each component is refined by matching a damped oscillation model;
- Correction of Gibbs effect and parameter estimation: the estimate of the full signal is further refined in order to remove Gibbs oscillations and to evaluate the frequency and damping coefficients of the oscillations.

In the following subsections these steps are described in detail.

### 4.1. Separation of Individual Oscillations

The Hilbert-based algorithm uses a decomposition method, presented and demonstrated in [21]. According to the theorem stated in [21], it is possible to describe the signal acquired by the PMU $y(t)$ as the sum of $m$ components $y_i^d(t)$:

$$\sum_{i=1}^{m} y_i^d(t). \tag{14}$$

The components $y_i^d$, with $i$ ranging from 1 to $m$, are such that their spectra $Y_i^d(f)$ are not overlapped and are significant only in a narrow frequency band, as show in Figure 1, delimited by the frequencies $f_{b_{i-1}}$ and $f_{b_i}$ (with $f_{b_0} = 0$), in the following referred to as bisecting frequencies; moreover, the sum of all the spectra $Y_i^d(f)$ is equal to the spectrum $Y(f)$ of the signal $y(t)$.

If $s_i(t)$ is the time domain signal corresponding to the portion of the spectrum $Y(f)$ of Figure 1 between the frequencies 0 and $f_{b_i}$, each of the components $y_i^d$ can be obtained by subtraction according to the following formulas:

$$y_i^d(t) = s_i(t) - s_{i-1}(t), \tag{15}$$

where, if $i$ is equal to 1, then $s_{i-1}(t) = s_0(t) = 0$ and if $i$ is equal to $m$, then $s_m(t) = y(t)$.

The terms $s_i(t)$ are evaluated, as demonstrated in [21] according to the Bedrosian theorem:

$$s_i(t) = \sin\left(2\pi f_{b_i} t\right) H\left[y(t)\cos\left(2\pi f_{b_i} t\right)\right] - \cos\left(2\pi f_{b_i} t\right) H\left[y(t)\sin\left(2\pi f_{b_i} t\right)\right] \tag{16}$$

It is worth noting that the truncation of the spectra in Figure 1 generates oscillations on the estimated signal $s_i(t)$, well known as Gibbs effect. At this aim, in Equation (12) authors preferred the use of Hilbert-Boche transform [24] that is able to reduce this undesired phenomenon.

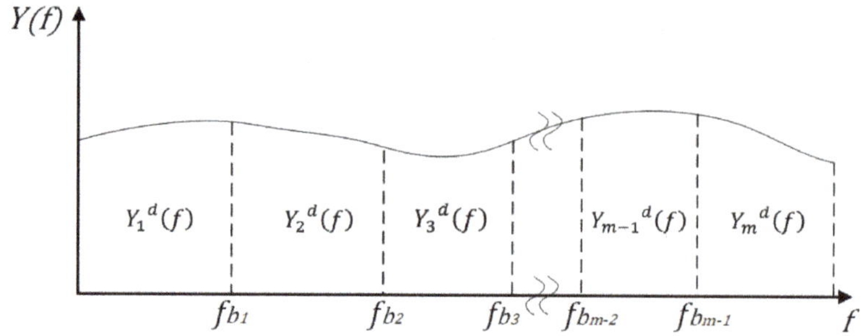


**Figure 1.** Spectra of the signal components.

The authors applied this method to extract the individual oscillations from the signal acquired by a PMU. It should be noted that, in this case, the spectra of the mono-components are overlapped, because the presence of damping causes the spread of the spectrum. The estimated components $y_i^d$ therefore do not exactly fit the oscillations composing the input signal. In other words, the portion $Y_i^d$ of the spectrum results from the super-position of two components. Therefore, the separation algorithm will return a $y_i^d$ component that differs from the mono-component oscillation; the more overlapping the spectra of the component oscillations, the higher the estimation error.

In [21] the LP-periodogram is used to identify the number of components and the values $f_{b_i}$ of the bisecting frequencies that delimit the spectrum intervals. The number of oscillations is given by the peaks of the spectrum, whereas the value of $f_{b_i}$ is calculated as the arithmetic average of the frequencies corresponding to the periodogram peak.

If the signal of interest $y(t)$ can be modeled as a sum of damped oscillations, each of them is characterized by a narrow band spectrum. Thus, the presented separation method can be exploited in order to obtain the individual oscillations.

The component signals $y_i^d$ therefore, are the inter-area oscillations affecting the bus voltages, whose frequency and damping coefficients have to be estimated.

The separation is applicable if a bisecting frequency can be recognized. Thus the proposed method requires the detection of at least two peaks in the spectrum. If only one peak is visible, because the acquired signal involves only one oscillation or because it involves oscillations whose frequency distance is lower than the spectral resolution, the method fails.

Also the presence of undesired frequency components, not related to inter-area oscillation, could put in crisis the separation method. In the paper, however, authors have taken into account only canonical signals observed in power system; the robustness to other type of components is going to be studied in future works.

## Optimal Choice of Bisecting Frequencies

As aforementioned, in order to separate the oscillations, the bisecting frequency is typically evaluated as the average between the frequencies of two sequential peaks of the periodogram of the input signal [21]. Actually, the optimal separation does not correspond to this frequency. For the sake of clarity, a signal obtained by the sum of two damped oscillations is taken into account. In particular, the signal is described by the following analytical expression:

$$y(t) = 30e^{-0.3t} \sin(\pi t) + 5e^{-0.1t} \sin(1.6\pi t). \quad (17)$$

The spectrum of this signal, shown in Figure 2 exhibits only two peaks centered at the frequencies 0.8 Hz and 0.5 Hz. The bisecting frequency, evaluated as average frequency, is equal to:

$$f_b = \frac{f_1 + f_2}{2} = \frac{0.8 + 0.5}{2} = 0.65 \text{ Hz} \tag{18}$$

The separation algorithm, described in the previous section, performed with $f_b$, provides two components, whose time evolution, compared with that of the nominal oscillations, is shown in Figure 3.

Besides an amplitude error caused by the Gibbs effect, it can be noted that even the zero crossing of the estimated components do not coincide with the oscillations composing the input signal.

This means that the estimation of the frequency of the two mono-component signals suffers from a slight deviation from the actual value.

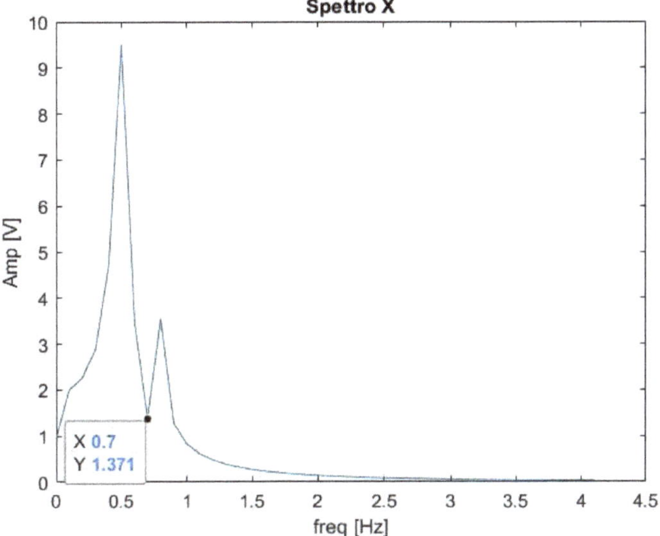


**Figure 2.** Spectrum of the signal $y(t)$ of Equation (17).

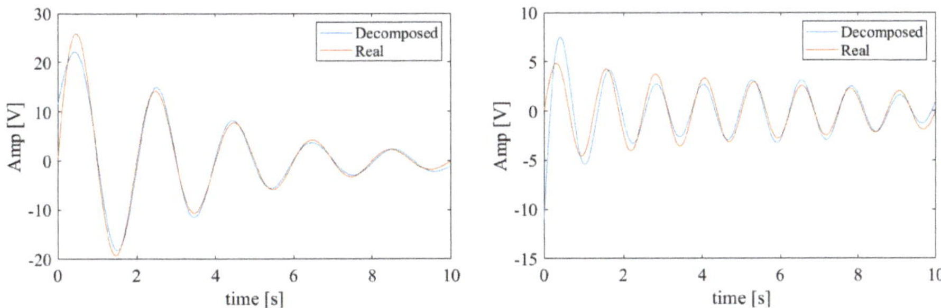


**Figure 3.** Mono-components obtained from the signal of Equation (17), setting the bisecting frequency equal to the average frequency; low frequency signal (**on the left**) and high frequency signal (**on the right**).

The authors, therefore, decided to improve the criterion for choosing the bisecting frequency. In particular, the frequency corresponding to the minimum of the portion of the signal spectrum delimited by two peaks is selected as the bisecting frequency.

In this way, in fact, the separation error is minimized when the spectra of the two components are partially overlapped.

As it can be appreciated from the spectrum of Figure 2, the estimated bisecting frequency, in this case, is equal to 0.7 Hz; the corresponding mono-component signals obtained by the separation algorithm are shown in Figure 4. It should be noted that, although the Gibbs effect is still present, the estimated oscillations match the frequency values characterizing the nominal input signals.

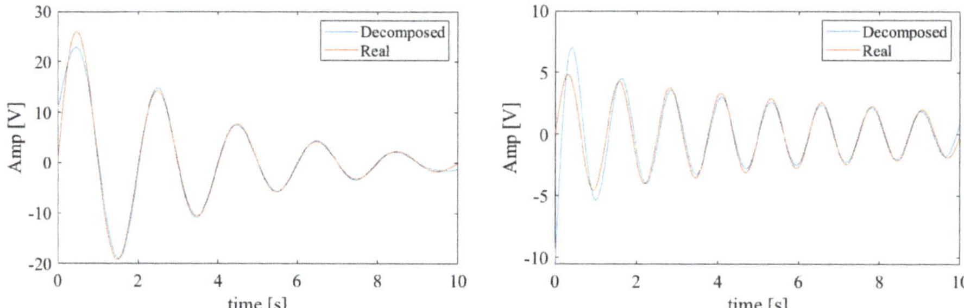


**Figure 4.** Mono-components obtained from the signal of Equation (17), setting the bisecting frequency equal to the minimum of the spectrum;low frequency signal (**on the left**) and high frequency signal (**on the right**).

### 4.2. Oscillations Reshaping

Despite of the adoption of some optimizations such as the Hilbert–Boche method and better choice of sectioning frequency, a deviation between the actual components and their estimates, due both to Gibbs effect and overlapping oscillations spectra, is still significantly appreciable. As noticeable in Figure 4, the estimate suffers from a distortion mainly evident at the oscillation edges, that hugely affects the damping estimate.

The second step of the proposed approach, then, relies on the reshaping of the mono-component signals. In detail, for each component, a non-linear least squares (NLS) regression algorithm is performed in order to detect the damped oscillation that is the best fit of the estimated mono-component.

At this aim the optimization tools of MATLAB® (R2018B, MathWorks, Natick, MA, USA) has been exploited [22], where the sequential quadratic programming (SQP) [25] has been set as minimization method and the fitting function is given by the formula:

$$y_i^{dr}(t) = A_i e^{d_i t} \sin(2\pi f_i t + \varphi_i), \quad (19)$$

where the unknown variables are the amplitude $A_i$, the damping coefficient $d_i$, the frequency $f_i$ and the starting phase $\varphi_i$; $y_i^{dr}(t)$ is the reshaped version of each estimated mono-component $y_i^d(t)$. Figure 5 shows the obtained oscillations after the reshaping performed on the mono-components of Figure 4. It can be noted that, for both the oscillations, the edge effects have been substantially reduced and the estimated frequencies almost match the actual ones. However, a slight amplitude deviation is still appreciable and, therefore, the oscillations estimate has to be further adjusted in order to properly estimate the damping coefficients.

The reshaped components are used as initial estimate for the successive step, that aims at accurately estimating the amplitude of the components and, consequently, the damping coefficient.

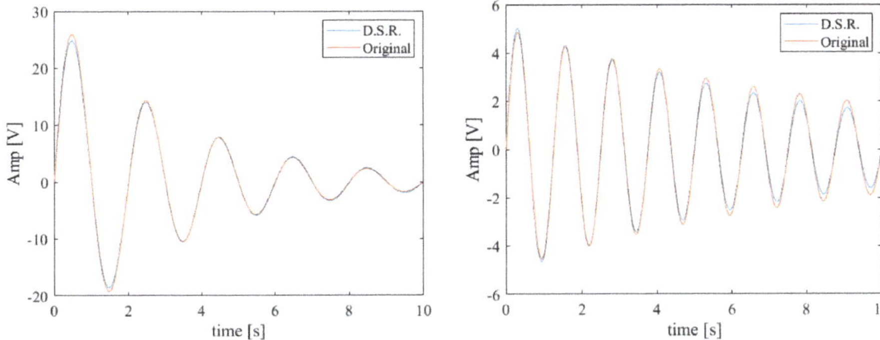


**Figure 5.** Mono-components obtained from the signal of Equation (17), after the reshaping process; low frequency signal (**on the left**) and high frequency signal (**on the right**).

### 4.3. Correction of Gibbs Effect and Parameters Estimation

Finally, the third step of the proposed approach consists of a further non-linear regression.

This time, the processed signal is the estimate of the overall signal $y_i^r(t)$, obtained by summing all the reshaped oscillations $y_i^{dr}(t)$.

Also this regression is an NLS, performed by exploiting the SQP method. However, the fitting function is given by:

$$\hat{y}(t) = \sum_{i=1}^{m} A_i e^{d_i t} \sin(2\pi f_i t + \varphi_i) \quad (20)$$

All the amplitudes, frequencies, damping coefficients and starting phases are the unknown variables. It should, however, be pointed out that the NLS algorithm converges rapidly (typically a time of less than 100 ms is required) since the initial values of the variables are those obtained from the second step and, therefore, already quite close to the values of the final fitting solution. Moreover, the NLS algorithm allows us to set the constraints (frequency and damping can vary in definite ranges) that limit the range of the possible solutions and favor the convergence.

In order to highlight the improvements introduced by the double regression algorithm, Table 1 shows the actual parameters and those estimated after the third step of the proposed method, for the components of the example signal described by equation Equation (17).

**Table 1.** Variable estimates after the third step of the proposed method.

| | **Component 1** | |
|---|---|---|
| | **Actual** | **Estimated** |
| Damping [$s^{-1}$] | 0.3 | 0.29999 |
| Frequency [Hz] | 0.5 | 0.49999 |
| Amplitude [V] | 30 | 29.99999 |
| | **Component 2** | |
| | **Actual** | **Estimated** |
| Damping [$s^{-1}$] | 0.1 | 0.09999 |
| Frequency [Hz] | 0.8 | 0.79999 |
| Amplitude [V] | 5 | 4.99999 |

## 5. Experimental Assessment

The proposed method has been assessed through both numerical and emulated experiments.

A preliminary design of the tests was required in order to set: (i) the duration of the signal time window to be analyzed; (ii) the sampling frequency characterizing the acquired signal, (iii) the characteristics of the test signal, and (iv) the figure of merit (FoM) to be used to evaluate the method performance.

The duration of the time window has been chosen in order to both minimize the response time of the algorithm and acquire at least one cycle of each mono-component signal (required condition for assuring the convergence of the NLS algorithm). The inter-area oscillations are characterized by a frequency range from 0.1 Hz up to 1 Hz; the window, assuring one cycle of a 0.1 Hz oscillation, is characterized by the width of 10 s.

As regards the sampling frequency, it has been selected according to the frequencies recommended for PMUs by Standard [26]. A minimum value of 5 S/s is mandatory to comply with the Nyquist–Shannon sampling theorem. However, low sampling rates correspond to few numbers of points acquired in each oscillation period,making unreliable the result of the NLS algorithms. On the other hand, higher sampling frequencies would result in increased number of points to be processed, and, consequently, higher computational burden and response time. A trade-off sampling frequency equal to 50 S/s has been selected. This sampling rate presents another advantage: being nominally equal to the fundamental frequency of the bus voltage, if the sampling operation is perfectly synchronized with the fundamental frequency, only the contribution of low frequency components is visible in the acquired signal.

The considered test signal is composed by two inter-area oscillations and is expressed by the formula:

$$y\left(t\right) = 1e^{-d_1 t}\sin\left(2\pi f_1 t\right) + 1e^{-d_2 t}\sin\left(2\pi f_2 t\right). \tag{21}$$

For the sake of simplicity, the amplitudes of both the oscillations have been set equal to 1 V and the starting phases has been set equal to zero. The frequency and the damping coefficient of the oscillations have been varied within their typical intervals reported in literature [1].

In particular, two sets of experiments have been planned. In the first one, referred in the following to as Test I, the frequencies $f_1$ and $f_2$ of the two oscillations have been set equal to 0.1 Hz and 0.6 Hz respectively. The damping coefficients, instead, are varied within the following intervals:

- $d_1 = [-0.05,\ \ 0.3]\ \mathrm{s}^{-1}$;
- $d_2 = [-0.05,\ \ 0.3]\ \mathrm{s}^{-1}$.

In the second set of experiments, referred in the following to as Test II, the damping coefficients have been set equal to 0.1 and 0.3 $\mathrm{s}^{-1}$, respectively, and the frequencies have been varied in order to change the frequency distance between the oscillations. In particular, an experiment has been performed by setting $f_1 = 0.1$ Hz and varying $f_2$ from 1 Hz to 0.3 Hz; another experiment has been carried out by setting $f_2$ equal to 1 Hz and varying $f_1$ from 0.1 to 0.8 Hz. This way, Thanks to the variation of the distance $|f_1 - f_2|$ between the frequencies of the two components, it has been possible to assess method performance in the case of components characterized by low (near 0.1 Hz) or high (near 1 Hz) oscillation frequency.

The selected FoM is the deviation between estimated $\hat{d}$ and nominal $d$ damping coefficients for both the oscillations:

$$\Delta d = \hat{d} - d \tag{22}$$

Particular attention, in fact, has been paid to the ability of the method to correctly estimate the oscillations damping coefficient, that is the parameter that mostly predicts potential system instability [27,28].

### *5.1. Numerical Tests with Ideal Signal*

The preliminary performance assessment of the proposed approach has been carried out through numerical tests in MATLAB® environment; to this aim, the signal of Equation (21) has been synthesized as input, according to the selected sampling rate of 50 S/s. The evolution of estimated values of FoMs versus the damping coefficients of the two mono-components for Test I is shown in Figure 6.

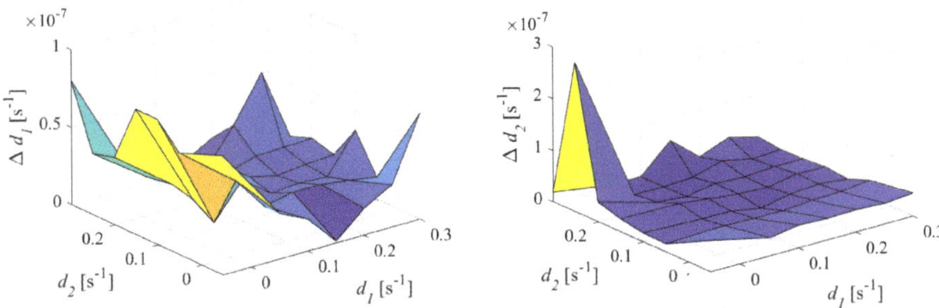


**Figure 6.** FoMs evolution versus damping coefficients obtained in numerical tests with ideal signal; results for frequency $f_1$ (**on the left**) and $f_2$ (**on the right**).

The proposed method exhibits excellent performance regardless the damping coefficients characterizing the oscillations. In fact, it is able to recognize and separate the two oscillations composing the test signal and, for both the components, it is able to estimate the damping coefficient with a $\Delta d$ values lower than $10^{-7}$ $\mathrm{s}^{-1}$ (corresponding to a relative percentage deviation of $10^{-4}$%).

The results obtained in test II conditions are shown in Figure 7. To better appreciate method performance, the logarithmic scale has been set for the vertical axis of both graphs.

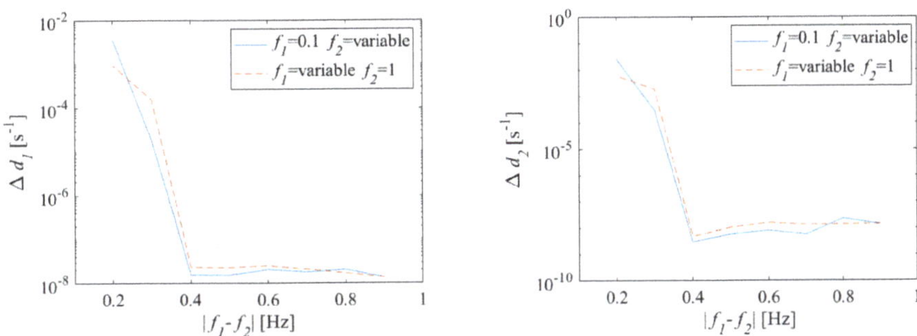


**Figure 7.** Figure of merit (FoM) evolution versus frequency distance between the components obtained in numerical tests with ideal signal; results for components with damping $d_1$ (**on the left**) and damping $d_2$ (**on the right**).

For both damping coefficients, estimate deviation $\Delta d$ was lower than $10^{-7}$ $\mathrm{s}^{-1}$ as long as the oscillation frequency distance was at least 0.4 Hz, corresponding to four times the spectral resolution associated with the sampling conditions. When $|f_1 - f_2|$ was reduced, an increase in the estimate deviation was experienced; however, when $|f_1 - f_2|$ is equal to 0.2 Hz, the estimated $\Delta d$ was lower than $10^{-2}$ $\mathrm{s}^{-1}$ for both the oscillations. This result was expected, since when the frequency are nearer, the spectra of the two mono-components are more overlapped and the distortion resulting from the separation procedure increases. A particular case consists of two oscillations whose frequency distance $|f_1 - f_2|$ was lower than 0.2 Hz; the method described in Section 4 failed, because the two oscillations were so near in frequency that a minimum in the signal spectrum is not detected and a bisecting frequency cannot be estimated. To obtain a suitable damping estimate even in such critical conditions, the measurement method has been improved by means of a further digital signal processing procedure that will be described in detail in Section 6.

Finally, $\Delta d$ turns out to be sensitive to the frequency distance between the two oscillations, but, given the same $|f_1 - f_2|$, the estimate of damping coefficients seemed not to sense if the oscillations were located at either high or low frequency.

### 5.2. Numerical Test with Noise and Quantization

Successive tests have been performed in conditions more similar to the actual sampling conditions. Thus, the effects of both noise and the quantization resulting from the analog to digital conversion stage have been added to the test signal of Equation (21).

The signal has been corrupted with white noise, whose amplitude has been set in order to assure a Signal to Noise Ratio (SNR) equal to 40 dB.

The use of an eight-bit analog-to-digital converter (ADC) has been hypothesized for introducing the quantization.

Since the presence of noise causes an inevitable dispersion of the algorithm results, the authors preferred to carry out $N$ equal to 50 repeated trials, for each frequency and damping condition. The damping estimates of the 50 tests were processed in order to evaluate the mean value and associated standard deviation. The FoM evaluated in these tests were the mean and the standard deviation of the measured $\Delta d_i$, with $i = 1, 2, \ldots, N$:

$$\mu\left(\Delta d\right) = \frac{1}{N}\sum_{i=1}^{N}\left(\Delta d_i\right) \tag{23}$$

$$\sigma\left(\Delta d\right) = \sqrt{\frac{1}{N-1}\sum_{i=1}^{N}\left(\Delta d_i\right)^2}. \tag{24}$$

Figure 8 shows the evolution of the estimated FoM versus damping coefficient, for both the oscillation in Test I configuration.

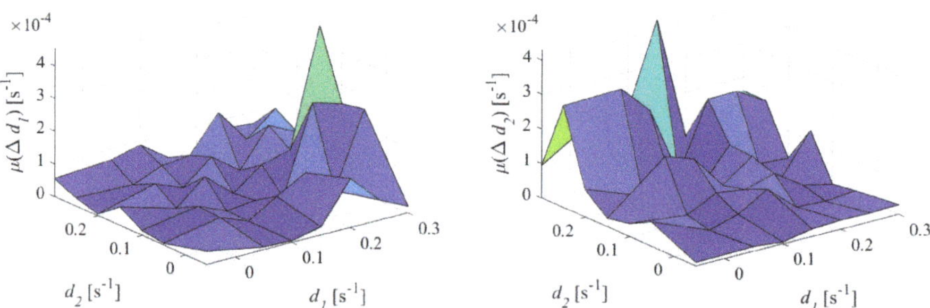


**Figure 8.** Mean of $\Delta d$ versus damping coefficients obtained in numerical tests with noise and quantization; results for frequency $f_1$ (**on the left**) and $f_2$ (**on the right**).

With respect to the values observed with ideal signals, an increase of estimate $\Delta d$ for both damping coefficients was appreciated. For both oscillations, however, the deviation of the estimate remained lower than $10^{-4}$ s$^{-1}$.

The standard deviations associated with the estimates of the damping coefficients are shown in Figure 9.

From the graph of the standard deviation associated with the damping estimates of the first oscillation, it can be noted that $\sigma(\Delta d_1)$ assumed values lower than $0.3 \times 10^{-3}$ s$^{-1}$ when the first oscillation was weakly damped. The higher the damping $d_1$, the higher the standard deviation, which reached up to about $2 \times 10^{-3}$ s$^{-1}$. The evolution of $\sigma(\Delta d_1)$ seemed to be independent on the values of $d_2$.

A similar behavior was observed on the damping estimates of the second oscillation: the standard deviation increased, up to $1.3 \times 10^{-3}$ s$^{-1}$, as the damping coefficient of the second oscillation rose, while it was not affected by the values of the damping coefficient of the first oscillation.

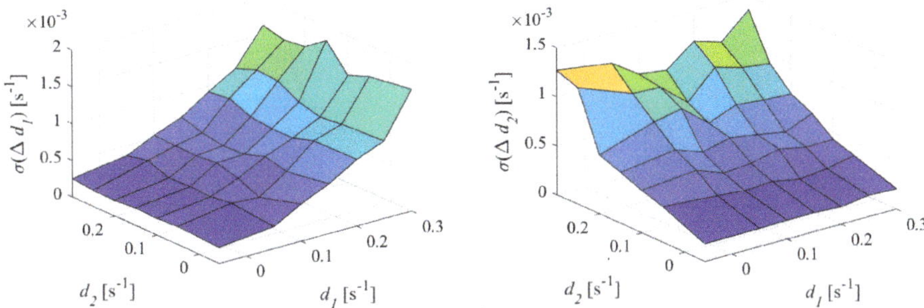


**Figure 9.** Standard deviation of $\Delta d$ versus damping coefficients obtained in numerical tests with noise and quantization; results for frequency $f_1$ (**on the left**) and $f_2$ (**on the right**).

It has to be noted that oscillations characterized by high damping coefficient are not dangerous for the system stability, therefore the behavior of the proposed method with respect to noise can be tolerated.

It can be concluded that the proposed algorithm was sensitive to the noise affecting the input signal and the sensitivity coefficient depends on the value of the damping coefficients. The noise, in particular, did not affect the mean of the damping estimates, but impacts directly on the dispersion of the results.

For the sake of clarity, the coefficient of variation (CV), proposed in [29] has also been evaluated. It is defined as:

$$CV = 100 \cdot \frac{\sigma_{\bar{d}_i}}{\bar{d}_i}, \tag{25}$$

where $\bar{d}_i$ is the mean value of the estimated $d_i$ and $\sigma_{\bar{d}_i}$ is the standard deviation of the mean. The obtained results are shown in Figure 10.

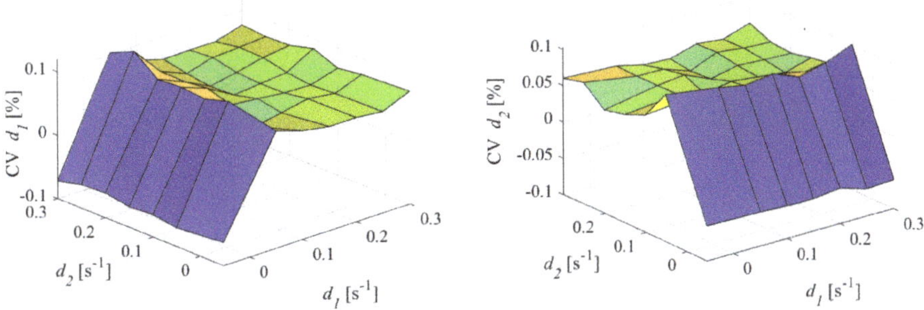


**Figure 10.** Coefficient of variation (CV) versus damping coefficients obtained in numerical tests with noise and quantization; results for frequency $f_1$ (**on the left**) and $f_2$ (**on the right**).

The absolute value of CV is always lower than 0.1%. The lower the damping coefficient $d_1$, the higher the CV; moreover, the CV of $d_1$ was not dependent on the values of $d_2$. Similar behavior can be observed for the CV of $d_2$. The results obtained from the experiments conducted according to Test II configuration are shown in Figure 11.

As long as $|f_1 - f_2|$ was higher than 0.3 Hz, measure $\Delta d$ was lower than $10^{-3}$ s$^{-1}$ for the first oscillation and lower than $10^{-2}$ s$^{-1}$ for the second oscillation. When $|f_1 - f_2|$ is 0.2 Hz, the estimate deviation increased for both damping coefficients as expected. The corresponding standard deviations are reported in Figure 12. As noticeable, the noise presence influenced the dispersion of the estimates of damping coefficient only when $|f_1 - f_2|$ is 0.2 Hz and, however, the standard deviation was lower than $10^{-3}$ $s^{-1}$.

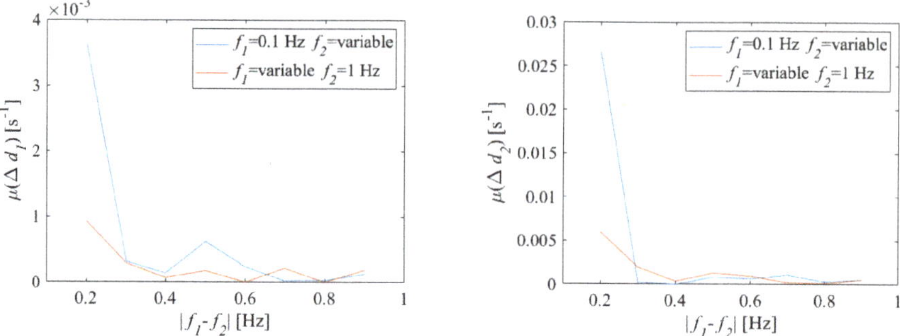


**Figure 11.** Mean of $\Delta d$ versus the frequency distance between the components obtained in numerical tests with noise and quantization; results for components with damping $d_1$ (**on the left**) and damping $d_2$ (**on the right**).

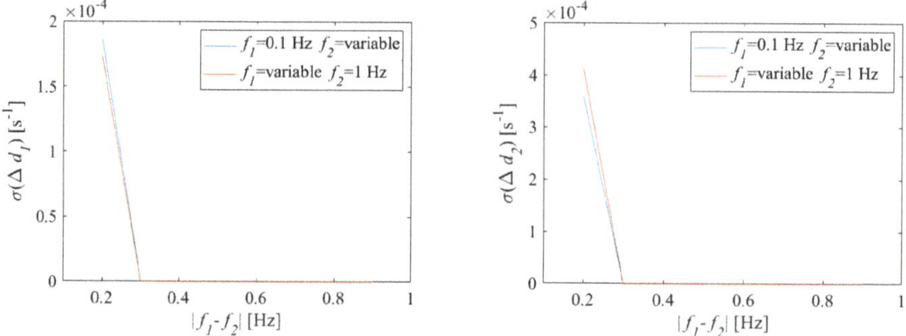


**Figure 12.** Standard deviation of $\Delta d$ versus the frequency distance between the components obtained in numerical tests with noise and quantization; results for components with damping $d_1$ (**on the left**) and damping $d_2$ (**on the right**).

### 5.3. Test Conducted on Digitized Signals

Finally, further tests have been performed on signals, emulating the composition of two oscillations, generated and acquired through real laboratory instrumentation.

To carry out the experimental tests, a laboratory test bench has been set up consisting of:

- Arbitrary waveform generator Agilent 33220A,
- Eight-bit digital oscilloscope Tektronix TDS 210,
- National Instruments GPIB-USB-HS interface.

A proper software has been developed in the National Instruments LabVIEW™ environment, in order to (i) synthesize the test signal according to Equation (21) in dependence on the desired frequencies and damping coefficients; (ii) transfer the waveform to the memory of the waveform generator, (iii) gather the samples acquired by the digital oscilloscope, and (iv) apply the proposed method to estimate the damping coefficients.

Again, 50 repeated measurements have been performed for each combination of damping coefficients and both the mean values and the standard deviations have been evaluated.

The mean of the deviation between the damping estimates and the nominal values is shown, for both oscillations, in Figure 13.

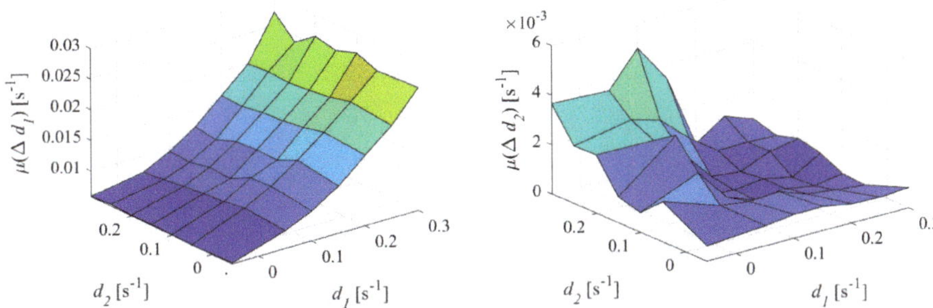


**Figure 13.** Mean of $\Delta d$ versus damping coefficients obtained in experimental tests with digitized signals; results for frequency $f_1$ (**on the left**) and $f_2$ (**on the right**).

It is evident that, in this case, the deviation of the damping estimates are characterized by values higher than those observed in numerical tests.

Moreover it can be noted that the deviation of the estimates exhibits a trend sensitive to the damping coefficients; in particular, the damping estimates worsen as the damping coefficient increases.

The evaluated standard deviations are shown in Figure 14.

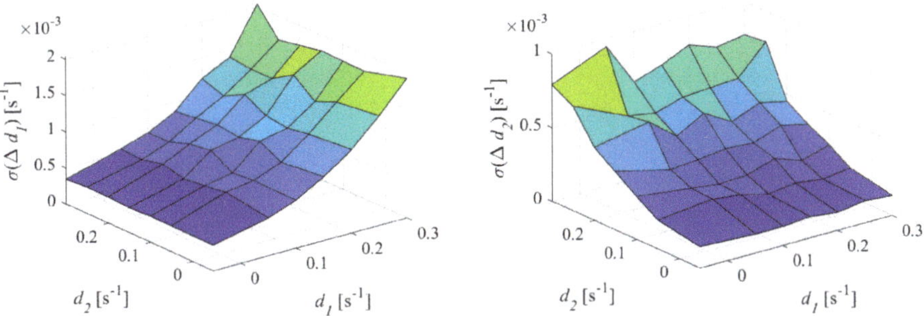


**Figure 14.** Standard deviation of $\Delta d$ versus damping coefficients obtained in experimental tests with digitized signals; results for frequency $f_1$ (**on the left**) and $f_2$ (**on the right**).

In this case, the trend and values of the standard deviation associated with the damping estimates is similar to those observed in numerical tests in the presence of noise and quantization presented in Section 5.2. The dispersion of the estimation results, therefore, is related exclusively to the presence of noise; the transition from numerical signals affected by a Gaussian noise to actually generated signals did not change the performance of the method, in terms of dispersion of measurement results.

The results obtained in Test II configuration shown in Figure 15.

Also in this case, it can be noted that the deviation of the mean value of the damping coefficient estimates has increased by about 10 times. For the sake of brevity the plots of the standard deviations and CV are not reported; however, values comparable with those observed in Section 5.2 have been appreciated.

It is worth noting that further tests have been carried out with oscillations characterized by different amplitude and with signal acquired through non-coherent sampling conditions. For the sake of brevity the graphs are not reported. However, the obtained results proved that also in these tests, the performance of the method are similar to the reported in this Section 5.2.

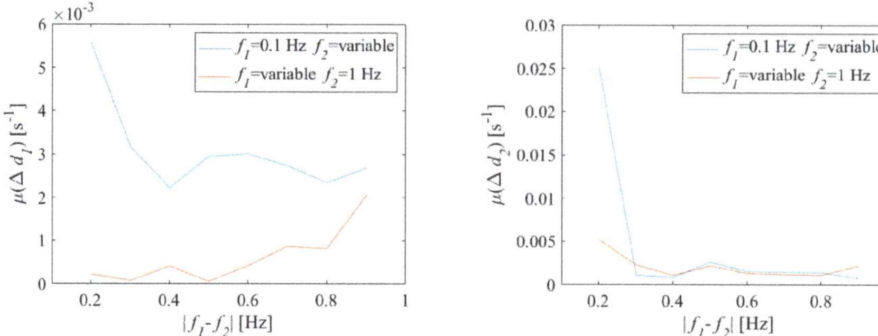


**Figure 15.** Mean of $\Delta d$ versus the frequency distance between the components obtained in experimental tests with digitized signals; results for components with damping $d_1$ (**on the left**) and damping $d_2$ (**on the right**).

### 5.4. Test Conducted on Inter-Area Oscillations Provided by Simulated Systems

In order to assess the method with a signal as similar as possible to a signal acquired during an inter-area oscillation, the Kundur's four-machine two-area test system [30], whose single line diagram is shown in Figure 16, has been adopted.

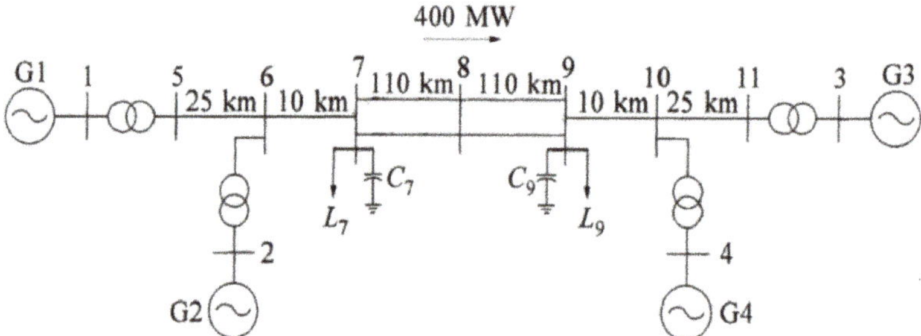


**Figure 16.** Single line diagram of Kundur's four machine two-area model adopted to obtain test inter-area oscillations.

The Kundur's model has been implemented in Simulink® (Simulink 9.2, MathWorks, Natick, MA, USA); a PMU to measure the bus voltage has been placed near the generator G2 and a three-phase fault has been simulated, in order to obtain the test signal. The voltage acquired by the PMU is nominally the combination of a local oscillation (the oscillation of G1 with respect to G2) and an inter-area oscillation (the oscillation of the area including G3 and G4 with respect to that including G1 and G2).

The initial part of the signal acquired by the PMU has been cut to remove the fault transient; the obtained signal, versus time, is shown in Figure 17, represented with the red dashed line.

The test signal has been processed according to the proposed approach. Two oscillations have been detected; in fact, the spectrum of the signal exhibited two peaks at the frequencies of 0.53 Hz (compatible with an inter-area oscillation) and 1.1 Hz (compatible with a local oscillation), respectively. The monocomponents estimated by the proposed method are shown in Figure 18.

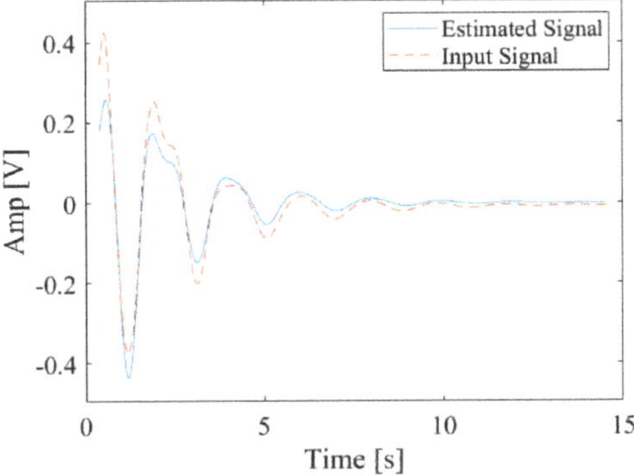


**Figure 17.** Test signal obtained by Kundur's Simulink® model and its estimation according to the proposed approach.

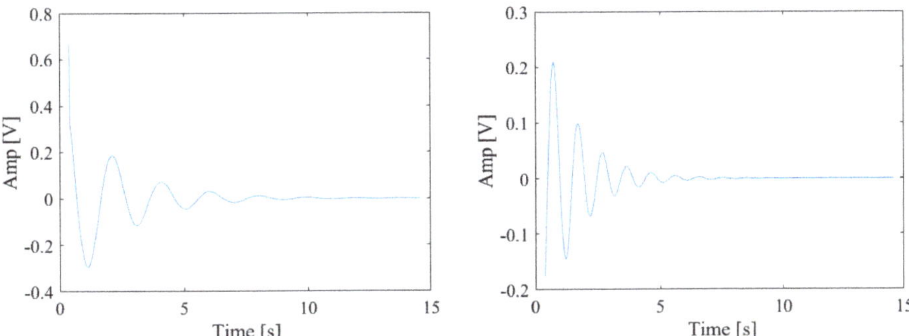


**Figure 18.** Individual oscillation estimated according to the proposed approach; low frequency signal (**on the left**) and high frequency signal (**on the right**).

In particular, the estimated parameters are reported in Table 2.

**Table 2.** Estimates of oscillations parameters of Kundir's model.

| Component | Frequency | Damping |
|---|---|---|
| Component 1 | 0.532 [Hz] | 0.475 [s$^{-1}$] |
| Component 2 | 1.121 [Hz] | 0.767 [s$^{-1}$] |

In order to evaluate the quality of the separation method, the two obtained monocomponents have been summed and the resulting signal, shown in Figure 17 and referred to as estimated signal, has been compared with the input test signal. It can be observed that the estimated signal accurately fits the time evolution of test signal.

## 6. Processing for Near Frequency Oscillations

As mentioned above, the proposed separation method fails if the oscillations are characterized by frequencies that are distant less than twice the spectral resolution.

The authors have improved the method, providing specific processing for this case, which is recognized because only one peak is detected in the spectrum.

In particular, the developed algorithm examines, by means of an NLS regression in the time domain, the signal acquired by the PMU through a moving time window; the algorithm stops when the root mean square (RMS) deviation between the input signal and the best fit is lower than a fixed threshold.

For the sake of clarity, the signal described by the following formula is considered:

$$y(t) = 1e^{0.05t}\sin(0.4\pi t) + 1e^{-0.3t}\sin(0.6\pi t). \tag{26}$$

The test signal consists of the sum of two oscillations, the first one divergent and the second one damped, characterized by frequencies equal to 0.2 Hz and 0.3 Hz, respectively. Both the time evolution and the spectrum of the test signal are shown in Figure 19.

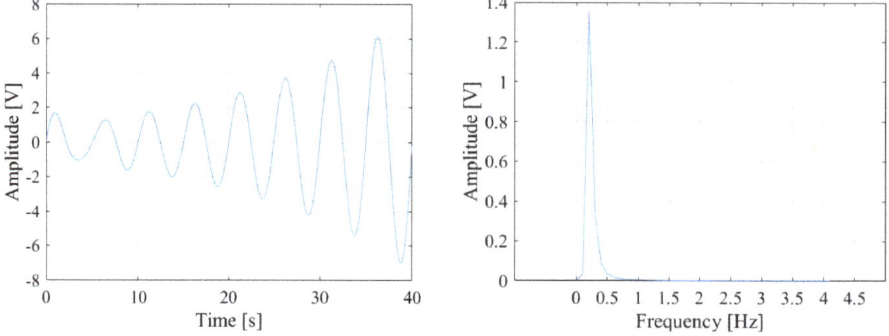


**Figure 19.** Time evolution (**on the left**) and spectrum (**on the right**) of the test signal adopted for the assessment.

It has to be observed that the spectrum of Figure 19 can also be associated to a single tone component, that has been acquired with non-coherent sampling conditions; so the separation of this signal in two oscillations could lead to unmeaningful information. For this reason, the algorithm search for the best mono-component oscillation that fits the input signal.

A 10 s time window of the test signal is shown in Figure 20. The samples of this window are given as input to an NLS minimization that fits the test signal with a single damped oscillation. If the actual signal is compared with the obtained best fit, it can be noted that the deviation in the estimate is high. The estimate of the damping coefficient is shown in the second column of Table 3.

**Table 3.** Estimates of the damping coefficient.

| Actual | Estimate on First Window | Estimate on Second Window |
|---|---|---|
| $-0.05$ [s$^{-1}$] | $-0.0103$ [s$^{-1}$] | $-0.0496$ [s$^{-1}$] |

This result is expected because, in the first 10 s, the analyzed signal is actually the combination of two oscillations, while the algorithm tries to fit the signal with only one oscillatory mono-component.

It should be noted, however, that although the NLS regression provides a rough estimate, it provides a negative sign of the damping. Therefore the regression is able to identify the presence of a divergent oscillation. This result can be exploited to raise an alert situation: the damping coefficient is not known accurately, but there is a divergent oscillation to keep under control.

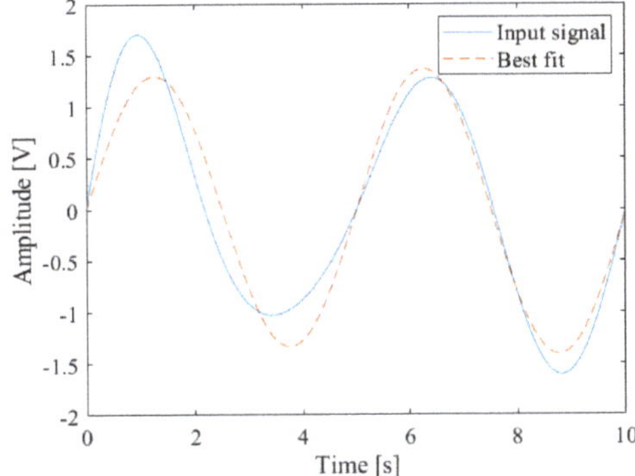


**Figure 20.** First 10 s time window of the acquired signal and its estimated best fit.

The next step in the algorithm is to examine the successive 10 s window and repeat the nonlinear regression. Both the signal and the best fit are shown in Figure 21. It can be noted that, this time, the best fit had a time evolution similar to that of the input signal. Damped oscillations, in fact, have been attenuated and the test signal is suitably approximable by a mono-component signal.

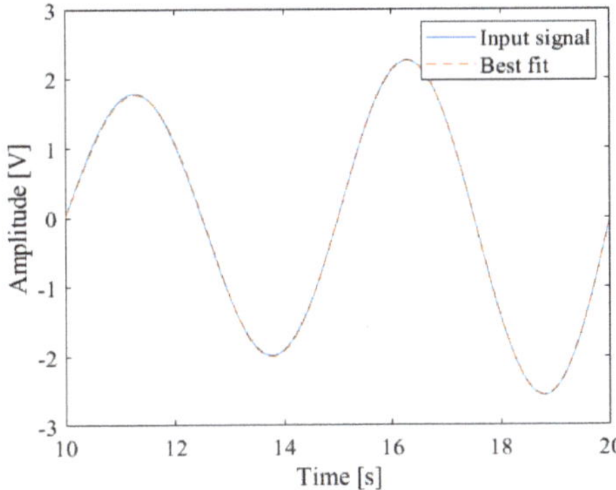


**Figure 21.** Second 10 s time window of the acquired signal and its estimated best fit.

The estimated damping is reported in the third column of Table 3.

The algorithm stops when the root mean square error (RMSE) between the samples of the acquired signal and those of the best fit is lower than a set threshold.

Summarizing, in the case of frequencies very close to each other, the proposed algorithm estimated a single damping, as if it detected only one oscillation. Various numerical tests have been performed, in different conditions, in order to verify the behavior of the algorithm. In the case of divergent oscillations, the algorithm is always capable of recognizing them since the estimated damping results negative in the first 10 s time window. Successively, as the time window moves, the algorithm refines the damping estimate.

Further tests have been carried out with signal involving only damped oscillations, also characterized by amplitudes quite different one from each other. In the first window, the algorithm can be strongly influenced by the component characterized by the highest amplitude; in this case the RMSE value can be very small, since the NLS result properly fits the acquired signal. The setting of the proper RMSE threshold, then, is a fundamental issue to assure the reliability of the method. A threshold equal to 0.1% of the amplitude has proved to be optimal in all the experiments.

It can be concluded that:

- if the signal contains a divergent oscillation, the estimated damping is the negative one, since the divergent oscillation exhibits a larger amplitude that masks the damped oscillations in the NLS minimization.
- if the signal contains only damped oscillations, the estimated damping coefficient is the one characterized by the smallest absolute value, i.e., that associated with the most persistent oscillation; therefore the algorithm is also able to recognize weakly damped oscillations.

## 7. Conclusions

In the paper a novel algorithm for the analysis and parameters estimation of inter-area oscillations has been proposed. The method consists of three steps: (i) the separation, according to an optimal bisection frequency, of the input signal in individual mono-component oscillations; (ii) reshaping of each mono-component in order to minimize the distortion due to the separation; (iii) correction of Gibbs effect and parameters estimation through a NLS algorithm.

Several tests, both with numerical and experimental signals have been carried out in order to assess the method. With respect to solutions already available in literature, the proposed method is characterized by excellent performance in terms of accuracy of the parameter estimates and robustness to noise.

As regards the computational burden and the response time, the proposed method allows to acquire and process a reduced number of samples, in favor of the computational resources. In particular, the described results have been obtained by processing 500 samples acquired with a sampling rate of 50 S/s. Concerning the response time, the time required by the algorithm execution is negligible with respect to the time spent for the acquisition of the signal samples. The method performance, in fact, is sensitive to the spectral resolution and, then, to the size of the analyzed time window. By considering the typical values of frequency and damping coefficient characterizing the inter-area oscillations, the authors have selected, as optimal trade off between the spectral resolution and the response time, a time window of 10 s.

The ongoing activity is focused on the development of a dedicated measurement instrument. At this aim, the realized algorithm has to be converted from MATLAB® and transferred on an embedded system.

**Author Contributions:** conceptualization, D.L. and A.L.; data curation, F.B., L.P.D.N. and S.T.; methodology, F.B., L.P.D.N., A.L. and S.T.; software, S.T.; supervision, D.L.; validation, F.B. and L.P.D.N.; writing—original draft, F.B., L.P.D.N. and S.T.; writing—review and editing, D.L. and A.L.

**Funding:** This research received no external funding.

**Acknowledgments:** The authors would like to thank Salvatore Lodato for his technical and scientific contribution to the research activity and to the drafting of the paper.

**Conflicts of Interest:** The authors declare no conflict of interest.

## References

1. Rogers, G. *Power System Oscillations*; Springer Science & Business Media: Berlin, Germany, 2012.
2. Jankowski, R.; Sobczak, B.; Trębski, R. Impact of the polish tie-lines' outages on the inter-area oscillations pattern in the synchronous system of continental Europe. *Acta Energetica* **2017**. 2017207. [CrossRef]
3. Zhang, S.; Xie, X.; Wu, J. WAMS-based detection and early-warning of low-frequency oscillations in large-scale power systems. *Electr. Power Syst. Res.* **2008**, *78*, 897–906. [CrossRef]
4. Prasertwong, K.; Mithulananthan, N.; Thakur, D. Understanding low-frequency oscillation in power systems. *Int. J. Electr. Eng. Educ.* **2010**, *47*, 248–262. [CrossRef]
5. Sauer, P.W.; Pai, M.A. *Power System Dynamics and Stability*; Pearson Prentice Hall: Upper Saddle River, NJ, USA, 1998; Volume 101.
6. Mohammadi-Ivatloo, B.; Shiroei, M.; Parniani, M. Online small signal stability analysis of multi-machine systems based on synchronized phasor measurements. *Electr. Power Syst. Res.* **2011**, *81*, 1887–1896. [CrossRef]

7. Semerow, A.; Höhn, S.; Luther, M.; Sattinger, W.; Abildgaard, H.; Garcia, A.D.; Giannuzzi, G. Dynamic Study Model for the interconnected power system of Continental Europe in different simulation tools. In Proceedings of the 2015 IEEE Eindhoven PowerTech, Eindhoven, The Netherlands, 29 June–2 July 2015; pp. 1–6.
8. Ma, J.; Han, D.; Sheng, W.J.; He, R.M.; Yue, C.Y.; Zhang, J. Wide area measurements-based model validation and its application. *IET Gener. Transm. Distrib.* **2008**, *2*, 906–916. [CrossRef]
9. Cai, G.; Yang, D.; Liu, C. Adaptive wide-area damping control scheme for smart grids with consideration of signal time delay. *Energies* **2013**, *6*, 4841–4858. [CrossRef]
10. Korba, P. Real-time monitoring of electromechanical oscillations in power systems: First findings. *IET Gener. Transm. Distrib.* **2007**, *1*, 80–88. [CrossRef]
11. Peng, J.C.H.; Nair, N.K.C. Enhancing Kalman filter for tracking ringdown electromechanical oscillations. *IEEE Trans. Power Syst.* **2011**, *27*, 1042–1050. [CrossRef]
12. Xiao, H.; Wei, J.; Liu, H.; Li, Q.; Shi, Y.; Zhang, T. Identification method for power system low-frequency oscillations based on improved VMD and Teager–Kaiser energy operator. *IET Gener. Transm. Distrib.* **2017**, *11*, 4096–4103. [CrossRef]
13. Hwang, J.K.; Liu, Y. Identification of interarea modes from ringdown data by curve-fitting in the frequency domain. *IEEE Trans. Power Syst.* **2016**, *32*, 842–851. [CrossRef]
14. Yuan, Z.; Xia, T.; Zhang, Y.; Chen, L.; Markham, P.N.; Gardner, R.M.; Liu, Y. Inter-area oscillation analysis using wide area voltage angle measurements from FNET. In Proceedings of the IEEE PES General Meeting, Providence, RI, USA, 25–29 July 2010; pp. 1–7.
15. Huang, N.E.; Shen, Z.; Long, S.R.; Wu, M.C.; Shih, H.H.; Zheng, Q.; Yen, N.C.; Tung, C.C.; Liu, H.H. The empirical mode decomposition and the Hilbert spectrum for nonlinear and non-stationary time series analysis. *Proc. R. Soc. Lond. Ser. A Math. Phys. Eng. Sci.* **1998**, *454*, 903–995. [CrossRef]
16. Baccigalupi, A.; Liccardo, A. The Huang Hilbert Transform for evaluating the instantaneous frequency evolution of transient signals in non-linear systems. *Measurement* **2016**, *86*, 1–13. [CrossRef]
17. Zhao, Y.; Li, Z.; Nie, Y. A time-frequency analysis method for low frequency oscillation signals using resonance-based sparse signal decomposition and a frequency slice wavelet transform. *Energies* **2016**, *9*, 151. [CrossRef]
18. Xiao, J.; Xie, X.; Han, Y.; Wu, J. Dynamic tracking of low-frequency oscillations with improved Prony method in wide-area measurement system. In Proceedings of the 2004 IEEE Power Engineering Society General Meeting, Denver, CO, USA, 6–10 June 2004; pp. 1104–1109.
19. Rai, S.; Lalani, D.; Nayak, S.K.; Jacob, T.; Tripathy, P. Estimation of low-frequency modes in power system using robust modified Prony. *IET Gener. Transm. Distrib.* **2016**, *10*, 1401–1409. [CrossRef]
20. Seppänen, J.M.; Turunen, J.; Haarla, L.C.; Koivisto, M.; Kishor, N. Analysis of electromechanical modes using multichannel Yule-Walker estimation of a multivariate autoregressive model. In Proceedings of the IEEE PES ISGT Europe 2013, Lyngby, Denmark, 6–9 October 2013; pp. 1–5.
21. Lauria, D.; Pisani, C. On Hilbert transform methods for low frequency oscillations detection. *IET Gener. Transm. Distrib.* **2014**, *8*, 1061–1074. [CrossRef]
22. Lauria, D.; Pisani, C. Improved non-linear least squares method for estimating the damping levels of electromechanical oscillations. *IET Gener. Transm. Distrib.* **2014**, *9*, 1–11. [CrossRef]
23. Bedrosian, E. A Product Theorem for Hilbert Transforms. *Proc. IRE* **1963**, *51*, 868. [CrossRef]
24. Boche, H.; Protzmann, M. A new algorithm for the reconstruction of bandlimited functions and their Hilbert transform. *IEEE Trans. Instrum. Meas.* **1997**, *46*, 442–444. [CrossRef]
25. Gill, P.E.; Wong, E. Sequential quadratic programming methods. In *Mixed Integer Nonlinear Programming*; Springer: Berlin, Germany, 2012; pp. 147–224.
26. IEEE Standard for Synchrophasor Data Transfer for Power Systems. *IEEE Std. C* **2011**, *37*, 1–53.
27. He, P.; Arefifar, S.A.; Li, C.; Wen, F.; Ji, Y.; Tao, Y. Enhancing oscillation damping in an interconnected power system with integrated wind farms using unified power flow controller. *Energies* **2019**, *12*, 322. [CrossRef]
28. Zuo, J.; Li, Y.; Shi, D.; Duan, X. Simultaneous robust coordinated damping control of power system stabilizers (PSSs), static var compensator (SVC) and doubly-fed induction generator power oscillation dampers (DFIG PODs) in multimachine power systems. *Energies* **2017**, *10*, 565.

29. de Souza, P.A., Jr.; Brasil, G.H. Assessing uncertainties in a simple and cheap experiment. *Eur. J. Phys.* **2009**, *30*, 615. [CrossRef]
30. Kundur, P.; Balu, N.J.; Lauby, M.G. *Power System Stability and Control*; McGraw-Hill Education: New York, NY, USA, 1994; Volume 7.

*Article*

# Accurate Assessment of Decoupled OLTC Transformers to Optimize the Operation of Low-Voltage Networks

**Álvaro Rodríguez del Nozal [1,*], Esther Romero-Ramos [2] and Ángel Luis Trigo-García [2]**

1 Departamento de Ingeniería, Universidad Loyola Andalucía, 41014 Seville, Spain
2 Department of Electrical Engineering, Universidad de Sevilla, 41092 Sevilla, Spain; eromero@us.es (E.R.-R.); trigoal@us.es (Á.L.T.-G.)
* Correspondence: arodriguez@uloyola.es

Received: 7 May 2019; Accepted: 3 June 2019; Published: 6 June 2019

**Abstract:** Voltage control in active distribution networks must adapt to the unbalanced nature of most of these systems, and this requirement becomes even more apparent at low voltage levels. The use of transformers with on-load tap changers is gaining popularity, and those that allow different tap positions for each of the three phases of the transformer are the most promising. This work tackles the exact approach to the voltage optimization problem of active low-voltage networks when transformers with on-load tap changers are available. A very rigorous approach to the electrical model of all the involved components is used, and common approaches proposed in the literature are avoided. The main aim of the paper is twofold: to demonstrate the importance of being very rigorous in the electrical modeling of all the components to operate in a secure and effective way and to show the greater effectiveness of the decoupled on-load tap changer over the usual on-load tap changer in the voltage regulation problem. A low-voltage benchmark network under different load and distributed generation scenarios is tested with the proposed exact optimal solution to demonstrate its feasibility.

**Keywords:** active distribution networks; voltage control; on-load tap changer transformers; low-voltage grids

---

## 1. Introduction

Low-voltage (LV) networks are generating increasing interest for a variety of reasons, such as the massive deployment of smart meters [1], the growing presence of distributed renewable generation (DG) [2], and important new components such as electric vehicles (EVs) and storage devices (SDs) [3]. The eruption of all these new actors has completely changed the approach to planning and operating voltage levels in light of two main facts. Firstly, new low-carbon technologies cause new planning and operational problems that were never taken into account when passive consumers were the only clients connected to this voltage level [4–6]. Secondly, a much greater extent of new and detailed information is available from smart meters [7], generating a very valuable starting point for tackling these new planning and operational challenges.

Power quality is one of the main concerns in an LV grid. Classical consumers and the aforementioned new participants demand quality from the distribution utility, with the level of voltage being one of the most important aspects. Many distribution companies have reported voltage complaints because of steady-state undervoltages and/or overvoltages [8–12].

On the one hand, permanent undervoltages/overvoltages are one of the issues that consumers are more concerned about. Both phenomena can lead to issues such as shutdowns, malfunctions, overheating, premature failures, and poor efficiency of consumer devices [13]. On the other hand, the level of supplied voltage is also of great relevance to the distribution company, and not only because of consumer complaints. It is well known that voltage control allows utilities to improve the operation efficiency of their networks by minimizing power losses [14], taking advantage of conservation voltage reduction [15], delaying new investments [16], and so on.

Some of the most common causes of undervoltage and overvoltage are the unsuitable tap setting of secondary distribution transformers, improper voltage regulation setting of substations, transformer/cable overloads (for undervoltages), and three-phase imbalances. As a general rule, the voltage of monitored networks tends to be closer to the upper voltage limits than the lower ones [11,12], which implies that the irruption of massive embedded generation can even worsen the frequency and duration of overvoltage phenomena [17].

The above details highlight that voltage control in low-voltage grids is of great relevance. Volt/var control of distribution networks has usually been confined to the medium-voltage (MV) level by using the on-load tap changer (OLTC) of high-voltage (HV)/MV transformers located at substations, capacitor banks, and/or step voltage regulators that are allocated along distribution feeders. Advanced volt/var control solutions specifically adapted to active distribution systems propose the use of new control sources: utility and customer-owned distributed power electronics (mainly those linked to DG), energy storage systems, power curtailment of DG generation and/or load, etc. All these novel control tools are well developed in the literature but not so much in practice [18,19]. For LV levels, in practice, most utilities with a European-style network design use only the secondary distribution transformers equipped with off-load tap changers to control voltages. Once again, the use of both the traditional and novel MV volt/var control solutions mentioned above has been intensively proposed in the most current literature, but it is not applied in practice [20–31]. The use of OLTC transformers in secondary distribution substations is currently the closest to realizing an LV reality [19,32–35]. This paper focuses on this last point: the use of OLTC MV/LV transformers (OLTCST) to control voltage in LV networks.

Depending on the technology, two different methods of operating an OLTCST are possible: (1) a synchronized tap change among the three phases or (2) decoupled control [32]. Most previous works have dealt with uniform and common tap positions for all three phases of the MV/LV-controlled transformer (3P-OLTCST) [19–21,28,30,31,33–35], but increasingly, more tap position settings among transformer phases are being proposed (1P-OLTCST) [24,25,29,32,36]. Decoupled control is obviously a great deal more attractive for LV networks because of their unbalanced nature.

Regardless of whether uniform or non-uniform phase tap movements are employed, there are two main methodologies for fulfilling voltage control: rule-based approaches or analytical solutions. Among the first group, different levels of "intelligence" have been proposed to reach a global solution that is as close as possible to the optimal one [21,24,28,32]. Analytical approaches are integrated solutions that solve an optimization problem to determine the best control actions that both minimize an objective function and fulfill all the electrical constraints [29,30,36]. While rule-based solutions reach a suboptimal solution but are easier to implement and need less network information, analytical approaches are more rigorous from a mathematical point of view and ensure not only an optimal solution but also that all the variables are within their limits. Until recently, the main drawbacks of applying analytical solutions laid in their computational cost (both programming and time consumed) and the necessity of having detailed models of the controlled network and scenarios of demand/generation for each time. Nowadays, the current trend toward active LV networks implies not only that these barriers are starting to disappear but also that the current knowledge of networks and power demand/dispersed generation can generate great value for distribution utilities [37].

On this basis, this paper tackles the optimal voltage control problem of European LV networks by using decoupled OLTC transformers (1P-OLTCST) and implementing an analytical solution. Previous works proposing analytical approaches to this problem have been mostly confined to American MV networks [30,36].

The authors in [30] use as voltage controllers 3P-OLTCST and PV inverters to minimize power losses, voltage violations, and lines loading; a three-phase simplified line model is assumed for four-wire three-phase lines as well as approximated voltage drop equations; no electrical model is taken into account for 3P-OLTCTS. The optimization problem proposed in [36] works with 1P-OLTCTS and static voltage converters, and although three-phase transformer models are incorporated into the mathematical problem, once again a three-phase simplified line model is considered. In addition, ground resistances different from zero are not taken into consideration. An optimal solution specifically designed for European MV and LV grids (both integrated) is published in [29]. In that approach, the optimal location and size of dSTATCOM, SD, and 3P-OLTCST, as well as the reactive contribution by DG's inverters are determined in order to manage voltage and minimize investment, operation, and maintenance costs. However, no details are provided in [29] to clearly deduce if detailed models for transformers, lines, and ground resistances are considered. In contrast, this new proposed work takes note of previous shortcomings and considers them warnings of the importance of neutral grounding considerations [38] and the need to use a rigorous model for all system components [39]. Therefore, the main novelties of this paper are the following:

- The optimization problem associated with the optimal voltage control of unbalanced LV networks by using transformers with OLTCs as control devices is rigorously defined and solved. No approximations are assumed and detailed models for all the components are considered:
  - three-phase model of three-phase transformers, incorporating also their ground connection design;
  - three-phase four-wire lines are neither reduced to a three-phase model nor decoupled,
  - any earthing system type is taken into account.
- The former optimization problem is implemented and solved, comparing the real possibilities of 1P-OLTCST versus 3P-OLTCST.

From the knowledge of the authors, no previous work has posed this voltage optimization problem for unbalanced LV networks with such a level of detail, so the results are quite conclusive.

The layout of the paper is as follows: Section 2 describes the framework in which voltage control is performed, and Section 3 tackles the mathematical problem definition by considering the appropriate model for each of the components entering the studied system. Section 4 shows the obtained results for a European low-voltage benchmark system to demonstrate both the benefits resulting from decoupled tap control compared with the traditional uniform three-phase tap control and the importance of the model to carry out effective control. Finally, Section 5 presents the most relevant conclusions and suggests the next steps for future work.

## 2. Low-Voltage European Networks

Distribution networks with the so-called European design are not limited to Europe; in fact, they are widely used all over the world. Many other countries adopted this arrangement, and the European design is well established, even in urban North American distribution networks [40]. The general structure of European distribution systems consists of a three-phase medium-voltage system in which three phase distribution transformers are connected to feed a large number of consumers/distributed generators connected to the three-phase low-voltage grid. An LV network can have a radial, ring, or meshed structure, but most are operated radially. Figure 1 depicts the schema of a typical radial European low-voltage grid.

Distribution or secondary transformers are large three-phase transformers (normally between 100 and 1000 kVA), with 20 kV/400 V being the most typical normalized voltages. One of the most common configurations is delta-wye because this arrangement prevents imbalance on the low-voltage side (due to unbalanced consumption) by transferring loads to the MV level. Nowadays, most of these secondary transformers have off-load tap changers, and OLTC transformers are considered in the rest of the paper, as justified in Section 1.

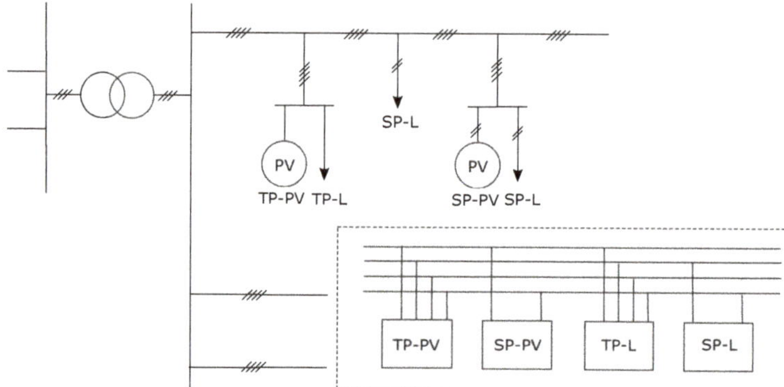


**Figure 1.** Schema of a common European low-voltage (LV) network.

Low-voltage lines are three-phase four-wire lines that are usually underground cables in urban areas and overhead lines in rural ones. LV networks are primarily earthed at the star-point of the MV/LV transformer. The earthing at the customer location is usually by the public supply network, and it is most often performed by earthing the neutral along different points of the low-voltage feeder.

Single-phase and three-phase consumers connect to the low-voltage network, with the former more prevalent than the latter. With distributed generators, mainly those that are photovoltaic, single-phase or three-phase consumers can also be connected, depending on the nominal power [41]. Single-phase clients connect one phase to the neutral conductor in an attempt to distribute them among the three phases so that the total three-phase equivalent net load from the secondary side of the transformer is as balanced as possible. In practice, although the same kind of single-phase consumers are equally distributed among all three phases, imbalances occur as a consequence of the different hourly load patterns. These imbalances are one of the main concerns of utility engineers.

One of the main conclusions resulting from the former design is that a three-phase model that takes into account all phases and the neutral must be considered, and a single-phase equivalent circuit is not valid for analyzing the whole power system. Even a decoupled single-phase circuit to study each of the three phases independently would lead to erroneous results [42].

## 3. Optimization Problem

The radial LV system under study comprises $N+1$ nodes (buses are numbered from 0). Without loss of generality, bus 0 applies to the MV side of the distribution transformer at the head of the whole LV grid. The set of three-phase buses, loads, and generators are denoted by B, L, and G, respectively. Note that the number of branches is equal to the number of buses minus one because of the radiality of the system.

The optimization problem to be solved involves determining the best value of the tap positions of each phase of a three-phase distribution transformer in order to optimize the operation of the whole system at a specific time $h$ while keeping all quantities within limits. These tap positions are the control variables of the problem. The dependent variables are phase–ground voltages at each node $i$, $\mathcal{U}_i^T = [\mathcal{U}_{iq}]^T = [\mathcal{U}_{ia}\ \mathcal{U}_{ib}\ \mathcal{U}_{ib}\ \mathcal{U}_{in}]^T$, phase current injections at each node $i$, $\mathcal{I}_i^T = [\mathcal{I}_{iq}]^T = [\mathcal{I}_{ia}\ \mathcal{I}_{ib}\ \mathcal{I}_{ib}\ \mathcal{I}_{in}]^T$, phase current flows through series branch $ij$, $\mathcal{I}_{ij}^T = [\mathcal{I}_{ij,q}]^T = [\mathcal{I}_{ij,a}\ \mathcal{I}_{ij,b}\ \mathcal{I}_{ij,c}\ \mathcal{I}_{ij,n}]^T$, and current $\mathcal{I}_{ig}$ through the ground resistor at bus $i$ ($i, j, k$ are indexes associated with buses; $(ij)$ denotes a series branch between buses $i$ and $j$; $p$ refers to each of the three phases $a, b, c$; the neutral phase is denoted by $n$; and the indexes $m, q$ denote any of the three phases as well as the neutral).

The next subsections detail the objective function and constraints associated with the optimization problem to be solved. Equality and inequality constraints result from modeling the electrical behavior of all the components and from considering all of the operational limits imposed by these components and the fulfillment of required standards. The phase domain and phase coordinates are considered to set out the problem, with computations being performed in physical units. In what follows, each complex equation that appears should be transformed into two equivalent real equations; this step is omitted throughout the paper.

### 3.1. Objective Function

The minimization of power losses is proposed as the main target objective. The power losses of the system can be formulated from a power balance point of view, so the following objective Function (1) results:

$$min \quad \Re(\mathcal{S}_{loss}) = min \quad \Re(\mathcal{U}_0^T \mathcal{I}_0^*) + \sum_{i=1}^{N+1} \sum_{p} \left[ \mathcal{S}_{ip}^G + \mathcal{S}_{ip}^L \right], \tag{1}$$

where $\mathcal{S}_{ip}^G$ refers to the generated specified power at phase $p$ of node $i$, and $\mathcal{S}_{ip}^L$ denotes the demanded complex power of a load at phase $p$ of node $i$. Note the net injected power by shunt elements connected to bus $i$ is $\mathcal{S}_{ip} = \mathcal{S}_{ip}^G + \mathcal{S}_{ip}^L$, so the generated power $\mathcal{S}_{ip}^G$ takes values greater than zero while the demanded load $\mathcal{S}_{ip}^L$ takes negative values. It is important to underline that the objective Function (1) considers all losses, namely power losses in the transformer, lines, and ground resistors.

The minimization of imbalances, voltage drops, or neutral currents are other common objective functions to consider. Power losses have commonly been chosen since they represent a well-defined single key index whose lowest value is usually linked to the best voltage profile. Other possible choices are those defined at a node or branch level, and such options necessitate the definition of a single index for any of them (for example, if the minimization of voltage drops was chosen, a possible voltage index to quantify this objective could be the addition of the deviation of each node voltage from its nominal value). In sum, an objective function that differs from the minimization of power losses could be set out.

### 3.2. Medium-Voltage Grid Equivalent

A three-phase Thévenin equivalent network is considered in the model of an MV grid that is upstream of the distribution transformer, as depicted in Figure 2. A rigid grounded wye three-phase balanced ideal voltage source $\mathcal{E}_g^T = [E_g\underline{|0} \ E_g\underline{|-120} \ E_g\underline{|120} \ 0]^T$ in series with a single impedance $\mathcal{Z}_g$ is assumed. These MV equivalent network parameters correspond with the actual operational medium voltage and the short-circuit impedance, respectively.

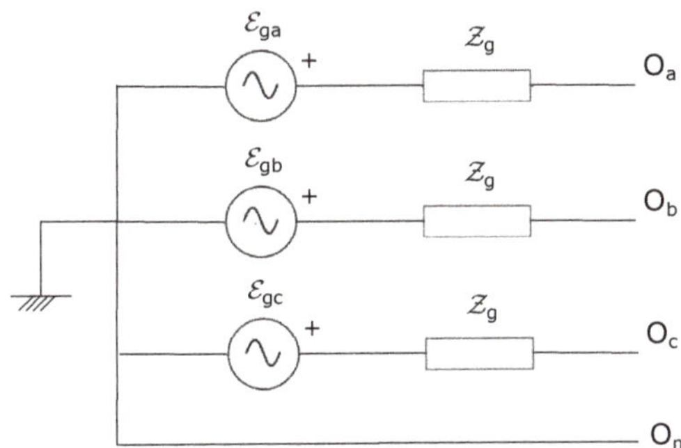


**Figure 2.** Thévenin equivalent for a medium-voltage (MV) grid.

### 3.3. Distribution Transformer Model

A complete and exact model for a three-phase transformer requires of a large number of short-circuit experiments, some of which are difficult to implement, depending on the type of transformer (common-core or shell-type integrated three-phase devices) [43]. A well-accepted starting assumption is that the three-phase transformer consists of three independent and identical single-phase transformers [43,44] since it has been demonstrated that the resulting inaccuracies from this simplification are subtle [43,45]. In this context, modeling the three-phase transformer starts by considering a single-phase transformer such as that shown in Figure 3, whose admittance-bus-based electrical model follows Equation (2):

$$\begin{bmatrix} \mathcal{I}_1 \\ \mathcal{I}_2 \end{bmatrix} = \begin{bmatrix} \mathcal{Y}_t + \mathcal{Y}_m & -r\mathcal{Y}_t \\ -r\mathcal{Y}_t & r^2\mathcal{Y}_t \end{bmatrix} \begin{bmatrix} \mathcal{U}_1 \\ \mathcal{U}_2 \end{bmatrix}, \tag{2}$$

where $\mathcal{Y}_t$ is the short-circuit or leakage admittance, $\mathcal{Y}_m$ is the open-circuit or magnetizing admittance—with both referring to the primary side of the transformer—and $r$ is the transformer ratio.

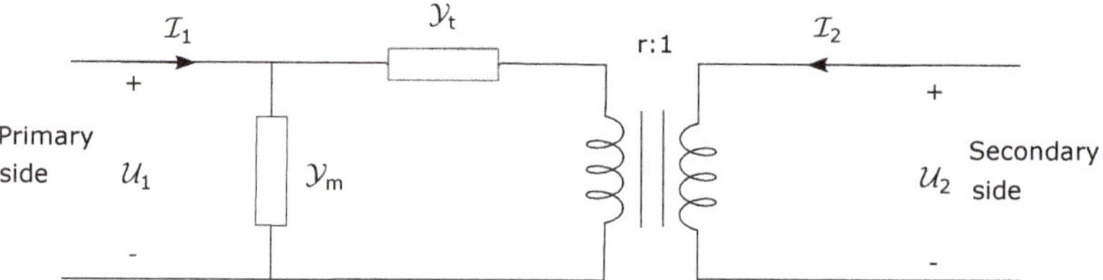


**Figure 3.** Equivalent circuit for a single-phase transformer.

The electrical model associated with three independent and decoupled single-phase transformers follows Equation (3), in which the three single-phase transformers are denoted by $a$, $b$, $c$, and their primary and secondary sides are numbered consecutively from 1 to 6. For further developments, three different transformer ratios are adopted for each of the individual phases and are denoted by $r_a$, $r_b$, and $r_c$.

$$\begin{bmatrix} \mathcal{I}_1 \\ \mathcal{I}_2 \\ \mathcal{I}_3 \\ \mathcal{I}_4 \\ \mathcal{I}_5 \\ \mathcal{I}_6 \end{bmatrix} = \left[\begin{array}{cc|cc|cc} \mathcal{Y}_{ta} + \mathcal{Y}_{ma} & -r_a\mathcal{Y}_{ta} & 0 & 0 & 0 & 0 \\ -r_a\mathcal{Y}_{ta} & r_a^2\mathcal{Y}_{ta} & 0 & 0 & 0 & 0 \\ \hline 0 & 0 & \mathcal{Y}_{tb} + \mathcal{Y}_{mb} & -r_b\mathcal{Y}_{tb} & 0 & 0 \\ 0 & 0 & -r_b\mathcal{Y}_{tb} & r_b^2\mathcal{Y}_{tb} & 0 & 0 \\ \hline 0 & 0 & 0 & 0 & \mathcal{Y}_{tc} + \mathcal{Y}_{mc} & -r_c\mathcal{Y}_{tc} \\ 0 & 0 & 0 & 0 & -r_c\mathcal{Y}_{tc} & r_c^2\mathcal{Y}_{tc} \end{array}\right] \begin{bmatrix} \mathcal{U}_1 \\ \mathcal{U}_2 \\ \mathcal{U}_3 \\ \mathcal{U}_4 \\ \mathcal{U}_5 \\ \mathcal{U}_6 \end{bmatrix} \tag{3}$$

Different three-phase transformers can result from these three independent single-phase transformers, depending on the connections between their terminal pairs, with wye and delta connections the most commonly used. The specific connection for the three-phase transformer allows for the formulation of mathematical expressions that link the voltages and currents of the three single-phase transformers with the voltages and currents of the high- and low-voltage sides of the three-phase transformer. For example, Figure 4 shows the $\Delta$−yg configuration, one of the most common schemes in European low-voltage networks. For this configuration, the relations among the former electrical quantities are defined by Equations (4) and (5), as can be easily deduced from Figure 4.

$$\begin{bmatrix} \mathcal{U}_1 \\ \mathcal{U}_2 \\ \mathcal{U}_3 \\ \mathcal{U}_4 \\ \mathcal{U}_5 \\ \mathcal{U}_6 \end{bmatrix} = \left[\begin{array}{ccc|cccc} 1 & -1 & 0 & 0 & 0 & 0 & 0 \\ 0 & 0 & 0 & 1 & 0 & 0 & -1 \\ 0 & 1 & -1 & 0 & 0 & 0 & 0 \\ \hline 0 & 0 & 0 & 0 & 1 & 0 & -1 \\ -1 & 0 & 1 & 0 & 0 & 0 & 0 \\ 0 & 0 & 0 & 0 & 0 & 1 & -1 \end{array}\right] \begin{bmatrix} \mathcal{U}_A \\ \mathcal{U}_B \\ \mathcal{U}_C \\ \mathcal{U}_a \\ \mathcal{U}_b \\ \mathcal{U}_c \\ \mathcal{U}_n \end{bmatrix} = N \begin{bmatrix} \mathcal{U}_A \\ \mathcal{U}_B \\ \mathcal{U}_C \\ \mathcal{U}_a \\ \mathcal{U}_b \\ \mathcal{U}_c \\ \mathcal{U}_n \end{bmatrix} \tag{4}$$

$$\begin{bmatrix} \mathcal{I}_A \\ \mathcal{I}_B \\ \mathcal{I}_C \\ \mathcal{I}_a \\ \mathcal{I}_b \\ \mathcal{I}_c \\ \mathcal{I}_n \end{bmatrix} = N^T \begin{bmatrix} \mathcal{I}_1 \\ \mathcal{I}_2 \\ \mathcal{I}_3 \\ \mathcal{I}_4 \\ \mathcal{I}_5 \\ \mathcal{I}_6 \end{bmatrix} + \begin{bmatrix} 0 \\ 0 \\ 0 \\ 0 \\ 0 \\ \frac{\mathcal{U}_n}{R_g} \end{bmatrix} \tag{5}$$

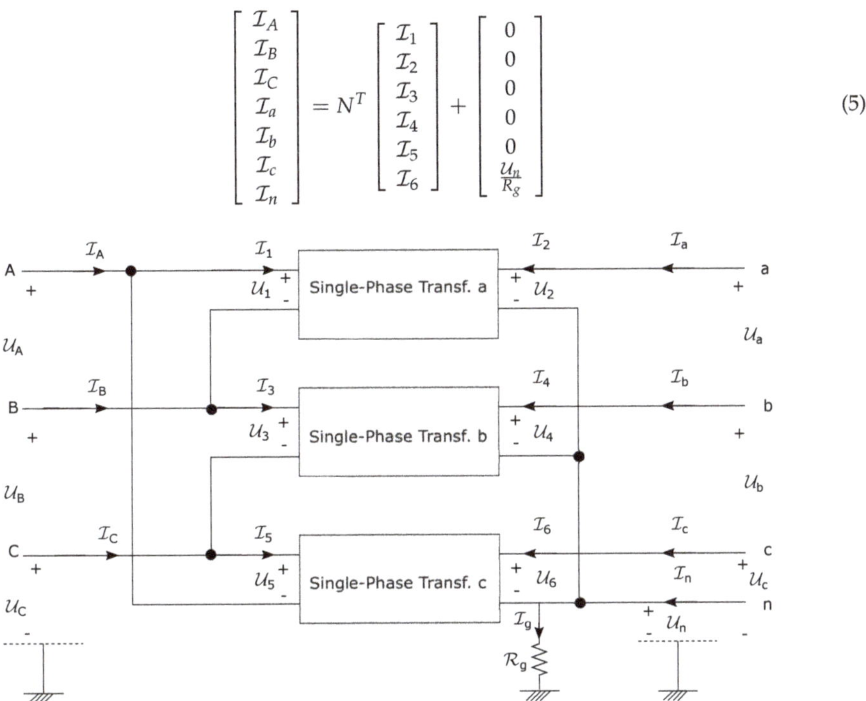


**Figure 4.** $\Delta-$yg three-phase transformer.

By substituting (3) into (5) and using (4), the final model (6) results for a three-phase $\Delta-$yg transformer that consists of three single-phase transformers ($\mathcal{Y}_{ta} = \mathcal{Y}_{tb} = \mathcal{Y}_{tc} = \mathcal{Y}_t$, $\mathcal{Y}_{ma} = \mathcal{Y}_{mb} = \mathcal{Y}_{mc} = \mathcal{Y}_m$) that are identical in all respects except for their transformer ratios:

$$\begin{bmatrix} \mathcal{I}_A \\ \mathcal{I}_B \\ \mathcal{I}_C \\ \mathcal{I}_a \\ \mathcal{I}_b \\ \mathcal{I}_c \\ \mathcal{I}_n \end{bmatrix} = \left[\begin{array}{ccc|ccc|c} 2\mathcal{Y}_t+2\mathcal{Y}_m & -\mathcal{Y}_t-\mathcal{Y}_m & -\mathcal{Y}_t-\mathcal{Y}_m & -r_a\mathcal{Y}_t & 0 & r_c\mathcal{Y}_t & (r_a-r_c)\mathcal{Y}_t \\ -\mathcal{Y}_t-\mathcal{Y}_m & 2\mathcal{Y}_t+2\mathcal{Y}_m & -\mathcal{Y}_t-\mathcal{Y}_m & r_a\mathcal{Y}_t & -r_b\mathcal{Y}_t & 0 & (r_b-r_a)\mathcal{Y}_t \\ -\mathcal{Y}_t-\mathcal{Y}_m & -\mathcal{Y}_t-\mathcal{Y}_m & 2\mathcal{Y}_t+2\mathcal{Y}_m & 0 & r_b\mathcal{Y}_t & -r_c\mathcal{Y}_t & (r_c-r_b)\mathcal{Y}_t \\ \hline -r_a\mathcal{Y}_t & r_a\mathcal{Y}_t & 0 & r_a^2\mathcal{Y}_t & 0 & 0 & -r_a^2\mathcal{Y}_t \\ 0 & -r_b\mathcal{Y}_t & r_b\mathcal{Y}_t & 0 & r_b^2\mathcal{Y}_t & 0 & -r_b^2\mathcal{Y}_t \\ r_c\mathcal{Y}_t & 0 & -r_c\mathcal{Y}_t & 0 & 0 & r_c^2\mathcal{Y}_t & -r_c^2\mathcal{Y}_t \\ \hline (r_a-r_c)\mathcal{Y}_t & (r_b-r_a)\mathcal{Y}_t & (r_c-r_b)\mathcal{Y}_t & -r_a^2\mathcal{Y}_t & -r_b^2\mathcal{Y}_t & -r_c^2\mathcal{Y}_t & \begin{array}{c}(r_a^2+r_b^2+r_c^2)\mathcal{Y}_t \\ +\frac{1}{R_g}\end{array} \end{array}\right] \begin{bmatrix} \mathcal{U}_A \\ \mathcal{U}_B \\ \mathcal{U}_C \\ \mathcal{U}_a \\ \mathcal{U}_b \\ \mathcal{U}_c \\ \mathcal{U}_n \end{bmatrix} \tag{6}$$

The described procedure allows the final bus admittance matrix to be obtained for any other group connection of the three transformers. Note that to obtain a perfect earth connection of the neutral, i.e., $R_g = 0$, the last diagonal element of the three-phase bus admittance matrix is not defined; in this case, $\mathcal{U}_n$ is zero and the last row of (6) is omitted. The neutral current flowing through the wye side of the transformer would be obtained by applying Kirchhoff's current law to the secondary bus of the transformer, as described in Section 3.6.

### 3.4. Low-Voltage Lines and Ground Resistances

Three-phase four-wire lines are the most common configuration for European low-voltage networks, and short single-phase sections are used to connect single-phase clients to the main three-phase four-wire LV line. The modeling of overhead and underground line segments must be as precise as possible since this model plays an important role in the final solution, as illustrated later. This high accuracy demands for a complete and reliable database in relation to low-voltage lines: the type and size of every conductor, the physical geometry of both overhead and underground lines, the resistance of the earth, etc. This information allows Carsons's equations [46] to be applied in order to obtain the resistance and the self- and mutual inductive reactance of the conductors that make up the line. These series impedances are shown in Figure 5. A 4 × 4 series impedance matrix $\mathcal{Z}_{ij}$ results from this model (7). The series impedance matrix is completely full for a three-phase four-wire line, with only non-null elements in one phase and neutral positions for two-phase or single-phase lines.

$$\mathcal{U}_i = \mathcal{Z}_{ij}\mathcal{I}_{ij} + \mathcal{U}_j = \begin{bmatrix} \mathcal{U}_{ia} \\ \mathcal{U}_{ib} \\ \mathcal{U}_{ic} \\ \mathcal{U}_{in} \end{bmatrix} = \begin{bmatrix} \mathcal{Z}_{ij}^{aa} & \mathcal{Z}_{ij}^{ab} & \mathcal{Z}_{ij}^{ac} & \mathcal{Z}_{ij}^{an} \\ \mathcal{Z}_{ij}^{ab} & \mathcal{Z}_{ij}^{bb} & \mathcal{Z}_{ij}^{bc} & \mathcal{Z}_{ij}^{bn} \\ \mathcal{Z}_{ij}^{ac} & \mathcal{Z}_{ij}^{bc} & \mathcal{Z}_{ij}^{cc} & \mathcal{Z}_{ij}^{cn} \\ \mathcal{Z}_{ij}^{an} & \mathcal{Z}_{ij}^{bn} & \mathcal{Z}_{ij}^{cn} & \mathcal{Z}_{ij}^{nn} \end{bmatrix} \begin{bmatrix} \mathcal{I}_{ij,a} \\ \mathcal{I}_{ij,b} \\ \mathcal{I}_{ij,c} \\ \mathcal{I}_{ij,n} \end{bmatrix} + \begin{bmatrix} \mathcal{U}_{ja} \\ \mathcal{U}_{jb} \\ \mathcal{U}_{jc} \\ \mathcal{U}_{jn} \end{bmatrix} \quad (7)$$

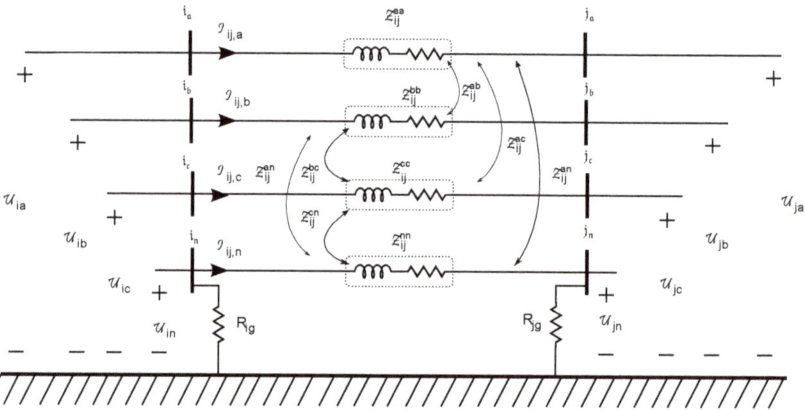

Figure 5. Three-phase four-wire electric line model.

It is common practice to neglect capacitances in the case of low-voltage lines because the lengths of the lines are short. So, throughout this paper, this shunt effect is not considered.

It is not always possible to apply Carson's equations because the requisite information is not usually available in utility-based data (distance between conductors, height of conductors, earth resistivity, etc.). Instead, positive sequence impedance $\mathcal{Z}_1$ and, more rarely, zero sequence impedance $\mathcal{Z}_0$ are the only practicable line data. In this last case, the best-approximated series impedance $\mathcal{Z}_{ij}^{Aprox}$ is

$$\mathcal{Z}_{ij}^{Aprox} = \begin{bmatrix} \mathcal{Z}_1 & 0 & 0 & 0 \\ 0 & \mathcal{Z}_1 & 0 & 0 \\ 0 & 0 & \mathcal{Z}_1 & 0 \\ 0 & 0 & 0 & \dfrac{\mathcal{Z}_0 - \mathcal{Z}_1}{3} \end{bmatrix}. \quad (8)$$

This simplified model is obtained by assuming an identical type of conductor for all phases and the neutral and equal sections, symmetrical spacing between phases (including the neutral), and perfect earthing at both extremes of each line segment [46].

Neutral conductors are commonly grounded at multiple points along the network (neutral grounding) for safety reasons. The values of these resistors depend on the type of earth conductor and electrode, as well as the resistivity of the terrain. These characteristics determine the value of the resistor $R_{ig}$ to consider at each bus $i$, and the value can ideally move between zero and infinity. If this resistor is a finite value, Ohm's law is applied (9):

$$\mathcal{U}_{in} = R_{ig}\mathcal{I}_{ig} \; \forall \, \text{bus } i \text{ earthed with a finite resistor,} \tag{9}$$

where $\mathcal{I}_{ig}$ denotes the current flowing through the ground resistor. Note that $\mathcal{U}_{in}$ has a null result for rigid earthing.

If the neutral of bus $i$ is not connected to the ground, that is, $R_{ig} = \infty$, the ground current is forced to be zero (10):

$$\mathcal{I}_{ig} = 0 \;\; \forall \, \text{bus } i \text{ unearthed} \tag{10}$$

### 3.5. Loads and Renewable Generators

Hereinafter, the appropriate model for consumers and renewable generators connected to the low-voltage system is discussed. The optimal power flow analysis discussed throughout this paper deals with the steady state, so static models are considered.

Static load models can be classified into the following groups: exponential, polynomial, linear, comprehensive, static induction motor, and power electronic-interfaced models [47]. Among all these static load models, the constant real and reactive power load model is the most widely used for utilities [47], as a particular case of the polynomial or ZIP model (constant impedance Z, constant current I, and constant power P model). Any other model could be considered if the appropriate parameters for each load are known. For renewable generation, with a focus on PV technologies, a steady-state power injection model is also adopted. PV generators operate by injecting the maximum available power, which varies depending on the irradiance. Similarly, the reactive power injection can be set depending on the standing grid code. Unity power factor operation is a common practice for PV generators connected to LV networks.

Three-phase loads and generators can be connected in wye or delta configurations, as shown in Figure 6. The neutral can be accessible or inaccessible in wye connections, and they are referred to as a three-phase four-wire connection or three-phase three-wire connection, respectively (see Figure 6a,b). The electrical model is given by (11) for a three-phase four-wire load/generator connected to bus $i$, while (12) is applicable to a three-phase four-wire load/generator.

$$\left.\begin{array}{l} \begin{bmatrix} \mathcal{S}_{ia}^{L/G} \\ \mathcal{S}_{ib}^{L/G} \\ \mathcal{S}_{ic}^{L/G} \end{bmatrix} = \begin{bmatrix} (\mathcal{U}_{ia} - \mathcal{U}_{in})(\mathcal{I}_{ia}^{L/G})^* \\ (\mathcal{U}_{ib} - \mathcal{U}_{in})(\mathcal{I}_{ib}^{L/G})^* \\ (\mathcal{U}_{ic} - \mathcal{U}_{in})(\mathcal{I}_{ic}^{L/G})^* \end{bmatrix} \\ \\ \mathcal{I}_{ia}^{L/G} + \mathcal{I}_{ib}^{L/G} + \mathcal{I}_{ic}^{L/G} + \mathcal{I}_{in}^{L/G} = 0 \end{array}\right\} \forall \text{ Three-phase four-wire - load L/ generator G} \in \text{bus } i \tag{11}$$

$$\left.\begin{array}{l} \begin{bmatrix} \mathcal{S}_{ia}^{L/G} \\ \mathcal{S}_{ib}^{L/G} \\ \mathcal{S}_{ic}^{L/G} \end{bmatrix} = \begin{bmatrix} (\mathcal{U}_{ia} - \mathcal{U}_{in_{L/G}})(\mathcal{I}_{ia}^{L/G})^* \\ (\mathcal{U}_{ib} - \mathcal{U}_{in_{L/G}})(\mathcal{I}_{ib}^{L/G})^* \\ (\mathcal{U}_{ic} - \mathcal{U}_{in_{L/G}})(\mathcal{I}_{ic}^{L/G})^* \end{bmatrix} \\ \\ \mathcal{I}_{ia}^{L/G} + \mathcal{I}_{ib}^{L/G} + \mathcal{I}_{ic}^{L/G} = 0 \end{array}\right\} \forall \text{ Three-phase three-wire wye- load L/ generator G} \in \text{bus } i, \tag{12}$$

where $L/G$ refers to the chosen load (L) or generator (G), depending on the considered component.

Equation (13) defines delta-connected three-phase loads or generators.

$$\left.\begin{array}{c}\begin{bmatrix} \mathcal{I}_{ia}^{L/G} \\ \mathcal{I}_{ib}^{L/G} \\ \mathcal{I}_{ic}^{L/G} \end{bmatrix} = \begin{bmatrix} \left(\dfrac{\mathcal{S}_{ia}^{L/G}}{(\mathcal{U}_{ia}-\mathcal{U}_{ib})}\right)^{*} - \left(\dfrac{\mathcal{S}_{ic}^{L/G}}{(\mathcal{U}_{ic}-\mathcal{U}_{ia})}\right)^{*} \\ -\left(\dfrac{\mathcal{S}_{ia}^{L/G}}{(\mathcal{U}_{ia}-\mathcal{U}_{ib})}\right)^{*} + \left(\dfrac{\mathcal{S}_{ib}^{L/G}}{(\mathcal{U}_{ib}-\mathcal{U}_{ic})}\right)^{*} \\ -\left(\dfrac{\mathcal{S}_{ib}^{L/G}}{(\mathcal{U}_{ib}-\mathcal{U}_{ic})}\right)^{*} + \left(\dfrac{\mathcal{S}_{ic}^{L/G}}{(\mathcal{U}_{ic}-\mathcal{U}_{ia})}\right)^{*} \end{bmatrix} \\ \mathcal{I}_{ia}^{L/G} + \mathcal{I}_{ib}^{L/G} + \mathcal{I}_{ic}^{L/G} = 0 \end{array}\right\} \forall\, \Delta\text{- load L/generator G} \in \text{ bus } i \quad (13)$$

It is worth remembering that $\mathcal{S}_{ip}^{L}$ values are negative and represent power consumed by loads, and $\mathcal{S}_{ip}^{G}$ values are positive and represent power generated by DGs.

Note that the same single-phase power has to be considered if a balanced three-phase load/DG is modeled ($\mathcal{S}_{ia}^{L/G} = \mathcal{S}_{ib}^{L/G} = \mathcal{S}_{ic}^{L/G}$). Single-phase loads and single-phase PVs are connected to one of the three phases $a, b, c$ and the neutral, or between two of the phases $a, b, c$. In these cases, the former equations apply, but null power/current injections have to be forced for the omitted phases.

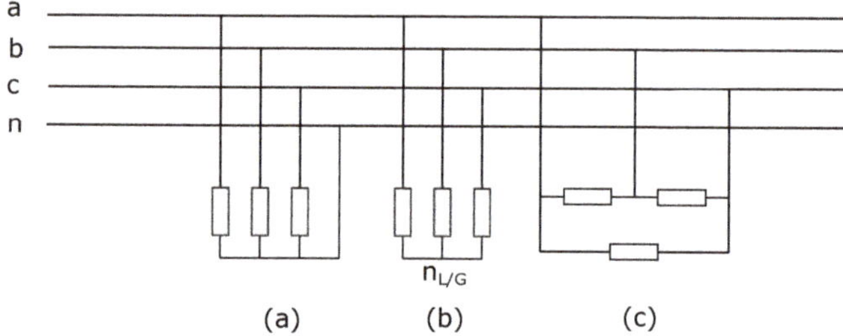


**Figure 6.** Example of three-phase wye (a and b) and delta (c) loads/renewable generators at bus $i$.

### 3.6. Kirchhoff's Laws

To complete the model of the system, Kirchhoff's laws have to be formulated. On one hand, Kirchhoff's voltage law has already been considered with Equation (7) when the voltage drop along a line is defined as the difference between the bus voltages of both line extremes; no additional equation is needed because of the radiality of the network. On the other hand, Kirchhoff's current law is easily developed by using the node incidence matrix $A$, which is associated with the oriented-type graph that describes the system topology. The number of rows in $A$ is equal to the number of series branches multiplied by 4 (because of the three phases and neutral), and the number of columns is four times the number of buses minus one (the reference bus is 0). Their elements are defined by (14). Series branches—transformers and lines—are always oriented from the reference bus towards the downstream.

$$A_{nodesxbranches} = \begin{cases} a_{k_m,(ij)_q} = +1 & \text{if phase m of bus } k \text{ is equal to phase } q \text{ of bus } i \\ a_{k_m,(ij)_q} = -1 & \text{if phase m of bus } k \text{ is equal to phase } q \text{ of bus } i \\ a_{k_m,(ij)_q} = 0 & \text{otherwise} \end{cases} \quad (14)$$

Once $A$ is obtained, Kirchhoff's current law is defined by (15):

$$A\left[\mathcal{I}_{ij,q}\right] = \left[\mathcal{I}_{km}\right], \quad (15)$$

where $\mathcal{I}_{km}$ is the net current injection entering bus $k$ through phase $m$ by any shunt component: loads, PV generators, and ground resistors. This leads to Equation (16):

$$\begin{array}{l} \left[\mathcal{I}_{kp}\right] = \sum_{L,G\in k}\left(\left[\mathcal{I}_{kp}^{G}\right] + \left[\mathcal{I}_{kp}^{L}\right]\right) \\ \mathcal{I}_{kn} = \mathcal{I}_{kn}^{G} + \mathcal{I}_{kn}^{L} - \mathcal{I}_{kg} \end{array} \quad \forall \, \text{bus}\, k \tag{16}$$

### 3.7. Operational Constraints

Secure operation of the system requires that voltage quantities remain within limits. This means that two inequality constraints are applied for each phase-neutral voltage at each bus. Therefore, upper and lower limits, $V^{\max}$ and $V^{\min}$, respectively, are imposed:

$$\left(V^{\min}\right)^2 \leq \left(V_{ip} - V_{in}\right)^2 \leq \left(V^{\max}\right)^2 \quad \forall \; i, \;\; p = a, b, c \tag{17}$$

Note that the constraint (17) is formulated in a quadratic form to avoid square roots, which cause strong nonlinearity in the resulting optimization problem.

## 4. Simulation Results

This section presents the results obtained for a benchmark European low-voltage network [48] when the optimal voltage control is achieved using 1P-OLTCST or 3P-OLTCST. The main objectives are to show the benefits of using decoupled tap control compared with traditional uniform three-phase tap control and illustrate the importance of the model in carrying out effective control.

The main aim of each optimization problem, 1P-OLTCST or 3P-OLTCST, is to provide a feasible solution regarding the fixed voltage bounds while minimizing the power losses of the system and fulfilling the power flow equations. If there is no solution that meets the voltage limit constraints, the solver provides the optimal solution without taking into account the lower voltage limit to avoid the non-convergence of the optimization problem.

This section is organized as follows: Section 4.1 presents the tested network and some considerations to take into account in the analysis; next, Sections 4.2 and 4.3 not only compare the two means of control (1P-OLTCST or 3P-OLTCST) for a specific load scenario but also demonstrate the importance of accuracy for the line model and earthing configuration. Finally, Section 4.4 incorporates distributed generation and compares both control strategies for a daily power scenario of both loads and generators.

### 4.1. Example Scenario Settings

In the subsequent simulations, let us consider the LV distribution network shown in Figure 7. In this grid, a three-phase medium-voltage system is connected to a three-phase low-voltage grid through the use of MV/LV transformers. The LV network has a radial structure in which three types of feeders are contemplated according to the connected consumers: residential, industrial, and commercial. Daily load profiles and detailed models of the different underground and overhead low-voltage lines considered in this work are provided in [48].

The following considerations apply in all the tested cases:

- In our study, a unique transformer MV/LV that feeds the three aforementioned subsystems is considered. The configuration of the transformer is $\Delta$−yg, and its nominal values are listed in Table 1. Short-circuit impedance $\mathcal{Z}_t$ refers to the MV primary side of the transformer.
- Slack bus, on the high-voltage side of the transformer, is assumed to be the nominal 20 kV, so an ideal three-phase wye voltage source is considered.
- A single tap is considered for each phase of the three-phase distribution transformer when decoupled tap control is applied (1P-OLTCST), while these three single taps move equally when the traditional three-phase control is considered (3P-OLTCST). Each tap has nine positions between the limits 0.92 and 1.08, in steps of 0.02.

- The nominal phase-neutral voltage is $V = 400/\sqrt{3}\,V$, and the Spanish normative is adopted with regard to voltage limits, that is, $\pm 7\%$ of $V_n$, which corresponds to a maximum phase-neutral voltage of 247.1 V and a minimum of 214.8 V.
- Loads and renewable generators are modeled as constant power, with a wye three-phase four-wire configuration in the case of three-phase control; single-phase loads are connected between one of the three phases $a, b, c$ and the neutral.
- All study cases were modeled using Python, and the stated optimization problem was solved with SciPy.

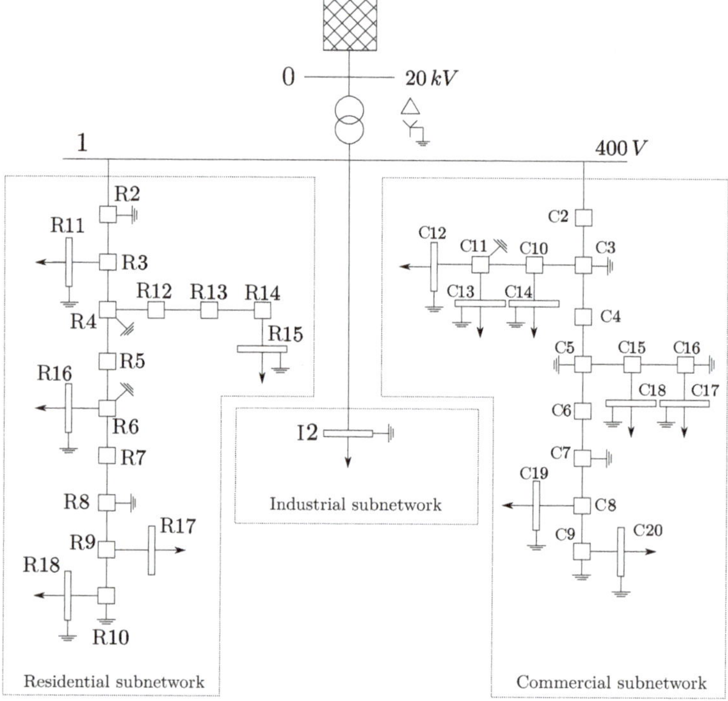


**Figure 7.** Modified European low-voltage benchmark network [48].

**Table 1.** Transformer parameters for the considered scenario.

| Bus from | Bus to | Connection | $\mathcal{V}_1$ [kV] | $\mathcal{V}_2$ [kV] | $\mathcal{Z}_t$ [Ω] | $S_{rated}$ [kVA] |
|---|---|---|---|---|---|---|
| 0 | 1 | Δ−yg | 20 | 0.4 | $21.164 + j69.616$ | 630 |

## 4.2. Line Modeling Accuracy

Let us consider (1) the load configuration detailed in Table 2 for a given time and (2) the neutral ground configuration shown in Table 3. Table 2 specifies the apparent load power demanded at each bus (negative values imply that power is not injected but consumed), power factor, and the percentage of the apparent power corresponding to each of the three phases and for every load bus. This unbalanced load scenario supposes that the total power demanded upstream is 319.85 kW and 118.78 kvar (without considering power losses) and that the proportions of the active power at each of the three phases $a$, $b$, and $c$ are 32.07%, 22.73%, and 44.2%, respectively. This scenario is not excessively imbalanced, although phase $c$ is more loaded, to the detriment of phase $b$.

Below, the optimization problem detailed in Section 3 is solved for the specified load scenario and earthing configuration. The two control strategies (1P-OLTCST and 3P-OLTCST) are not only considered but also solved if the exact line model provided in [48] is assumed (7) or if there are insufficient line data and a simplified model (s.m.) is the only possibility (8). Table 4 gives the simulation results for all four cases, as well as the base case corresponding to the nominal position of the taps, and once again, an exact line model or the approximated line model is used. The main figures are the phase transformer ratio, total active power losses (and savings in losses in percentage relative to the base case), and maximum and minimum phase-neutral voltages in the system (red figures are used to emphasize quantities out of limits). For those voltage controls assuming a simplified model, two quantities are reported for each magnitude: that obtained by the approximated model and the equivalent real value linked to the exact model (this last one between brackets), both computed with the optimal tap positions obtained by the corresponding optimization problem. For each type of transformer, the optimal solutions obtained are not the same when considering an accurate or a simplified model of the power lines. This fact makes the solver provide two different results for the transformer taps.

**Table 2.** Load configuration.

| **Bus** | **R2** | **R11** | **R15** | **R16** | **R17** | **R18** | **I2** | **C2** |
|---|---|---|---|---|---|---|---|---|
| $S$ [$kVA$] | −12 | −55 | −25 | −100 | −44 | −10 | −15 | −10 |
| $\cos\varphi$ | 0.95 | 0.95 | 0.95 | 0.95 | 0.95 | 0.95 | 0.85 | 0.95 |
| Ph. a (%) | 20 | 30 | 40 | 40 | 35 | 40 | 0 | 0 |
| Ph. b (%) | 20 | 20 | 30 | 10 | 25 | 20 | 70 | 70 |
| Ph. c (%) | 60 | 50 | 30 | 50 | 40 | 40 | 30 | 30 |
| **Bus** | **C8** | **C12** | **C13** | **C14** | **C17** | **C18** | **C19** | **C20** |
| $S$ [$kVA$] | −8 | −5 | −20 | −3 | −12 | −7 | −10 | −6 |
| $\cos\varphi$ | 0.90 | 0.90 | 0.90 | 0.90 | 0.90 | 0.90 | 0.90 | 0.90 |
| Ph. a (%) | 20 | 20 | 30 | 20 | 50 | 25 | 20 | 30 |
| Ph. b (%) | 10 | 20 | 40 | 45 | 20 | 0 | 30 | 10 |
| Ph. c (%) | 70 | 60 | 30 | 35 | 30 | 75 | 50 | 60 |

**Table 3.** Neutral ground configuration.

| **Bus** | **1** | **R4** | **R6** | **R8** | **R10** | **R11** | **R15** | **R16** | **R17** | **R18** | **I2** | **C3** |
|---|---|---|---|---|---|---|---|---|---|---|---|---|
| $R_g(\Omega)$ | 3 | 40 | 40 | 40 | 40 | 40 | 40 | 40 | 40 | 40 | 40 | 40 |
| **Bus** | **C5** | **C7** | **C9** | **C11** | **C12** | **C13** | **C14** | **C16** | **C17** | **C18** | **C19** | **C20** |
| $R_g(\Omega)$ | 40 | 40 | 40 | 40 | 40 | 40 | 40 | 40 | 40 | 40 | 40 | 40 |

It is worth pointing out that the 3P-OLTCST is not able to provide a feasible solution because the voltages of some buses under the lower limit are fixed in the problem formulation. However, the same case is efficiently solved by using a 1P-OLTCST that, relying on an independent tap position for each phase, maintains the voltage of all the buses within limits, in addition to notably improving the power losses, reducing them by almost 15%, which is 6 points more than when using the 3P-OLTCST. We emphasize the need for rigorousness in the model to be solved since it is evident from Table 4 that simplified line models fail to accurately comply with voltage constraints (note that maximum voltages are greater than the upper limit of 247.1). Moreover, simplified models prevent obtaining an optimal solution for tap positions, regardless of whether a three-phase OLTC or a single-phase OLTC for each phase is considered.

Figure 8 shows the phase-neutral voltages and neutral voltage at all buses for the 1P-OLTCST for both an accurate model and a simplified model of power lines. Figure 9 is the counterpart of Figure 8 for the 3P-OLTCST. It is easily deduced by their comparison that the 1P-OLTCST phase-neutral voltages become more balanced than those reached by the 3P-OLTCST, and the maximum voltage drop is quite a bit lower in the first case, with values of 23.3 V and 32.6 V at the residential feeder for

1P-OLTCST and 3P-OLTCST, respectively. This result is expected because of the improved control flexibility of 1P-OLTCST compared with 3P-OLTCST.

**Table 4.** Simulation results when the primitive impedance matrices or the simplified model (s.m.) are considered in the system line modeling.

| | Base (s.m.) | Base | 3P-OLTCST (s.m.) | 3P-OLTCST | 1P-OLTCST (s.m.) | 1P-OLTCST |
|---|---|---|---|---|---|---|
| $r_a$ | | | | | 0.94 | 0.94 |
| $r_b$ | 1 | 1 | 0.94 | 0.96 | 0.94 | 0.96 |
| $r_c$ | | | | | 0.92 | 0.92 |
| $P_{loss}$ $(W)$ | 21,598.7 | 22,985.3 | 18,961.6 (20,164.7) | 21,070.3 | 18,406.5 (19,545.7) | 19,554.5 |
| Losses (%) | | 100 | | 91.66 | | 85.07 |
| $V_{max}$ $(V)$ | 230.5 | 234.5 | 244.9 (248.7) | 243.8 | 246.3 (248.5) | 246.4 |
| $V_{min}$ $(V)$ | 197.7 | 200.3 | 214.4 (216.8) | 211.1 | 220.5 (223.0) | 223.1 |

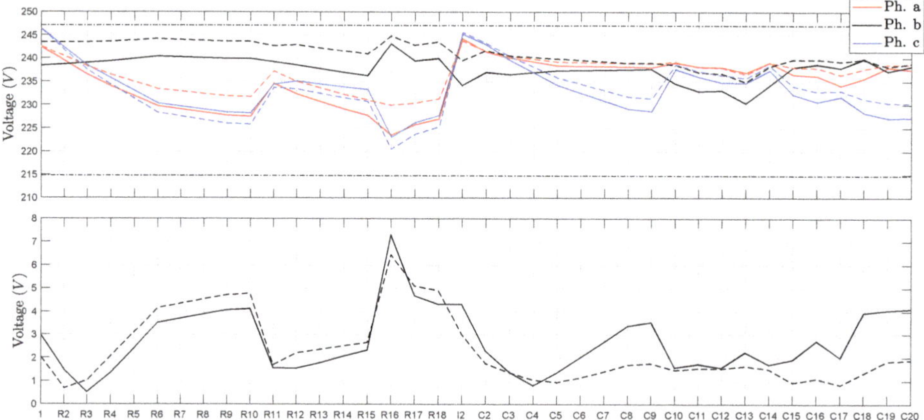


**Figure 8.** Simulation results when considering a 1P-OLTCST for both the accurate model and simplified model (dashed lines) of power lines.

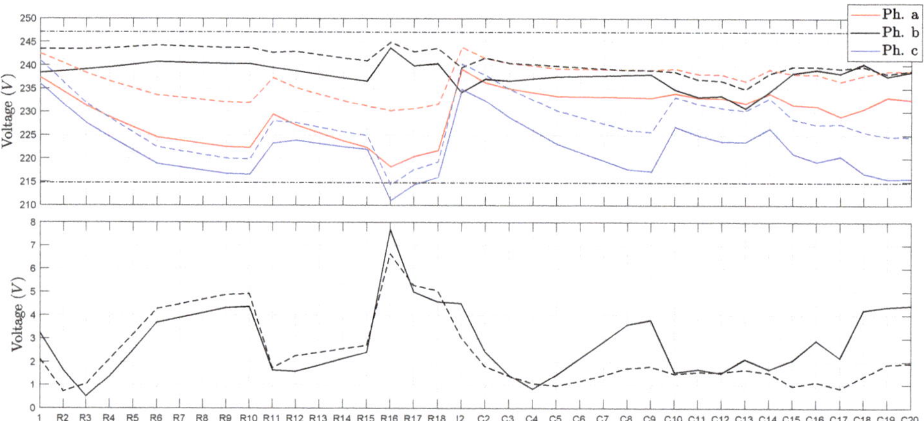


**Figure 9.** Simulation results when considering a 3P-OLTCST for both the accurate model and simplified model (dashed lines) of power lines.

### 4.3. Ground Configuration

In this second case study, the effects of the neutral ground configuration on the optimal tap positions, system losses, and network voltage values are shown. Let us consider the same load configuration as in the previous section (Table 2) but with different ground resistances.

**Table 5.** Neutral ground resistances at each bus.

| **Bus** | **1** | **R4** | **R6** | **R8** | **R10** | **R11** | **R15** | **R16** | **R17** | **R18** | **I2** | **C3** |
|---|---|---|---|---|---|---|---|---|---|---|---|---|
| $R_g(\Omega)$ | 3 | $R_1$ | $R_2$ | $R_1$ | $R_2$ | $R_1$ | $R_2$ | $R_1$ | $R_2$ | $R_1$ | $R_1$ | $R_2$ |
| **Bus** | **C5** | **C7** | **C9** | **C11** | **C12** | **C13** | **C14** | **C16** | **C17** | **C18** | **C19** | **C20** |
| $R_g(\Omega)$ | $R_1$ | $R_2$ | $R_1$ | $R_2$ | $R_1$ | $R_2$ | $R_1$ | $R_2$ | $R_1$ | $R_2$ | $R_1$ | $R_2$ |

From Table 5, the sequel ground configurations are analyzed:

- Case 1: Neutral resistor of finite value: $R_1 = R_2 = 40\,\Omega$.
- Case 2: Neutral rigid earthing: $R_1 = R_2 = 0\,\Omega$.
- Case 3: Neutral resistor of finite value: $R_1 = R_2 = 3\,\Omega$.
- Case 4: Mixed earthing: $R_1 = 40\,\Omega$, $R_2 = 0\,\Omega$.

Table 6 shows the simulation results for the four cases and for both types of control transformers, as well as the solutions for each case when the tap positions remain at their nominal values, for comparison purposes. We highlight that the optimal transformer taps are the same for all cases, except for the one in which a rigid earthing of the neutral is considered (Case 2), and for both types of control transformers. It can be seen that the results for a neutral grounding with a resistor of $3\,\Omega$ (Case 3) and a resistor of $40\,\Omega$ (Case 1) are very similar. However, only the 1P-OLTCST is able to maintain the voltage within the prescribed limits since the 3P-OLTCST is not able to satisfy voltage limits for any non-null value of grounding resistors. It is also noticeable that the 1P-OLTCST improves the power losses by around a 6% compared with the 3P-OLTCST in almost all cases. This improvement is reduced to 2.5% in the case of rigid grounding.

**Table 6.** Simulation results for different neutral ground configurations.

| | **1P-OLTCST** | | | | | | | |
|---|---|---|---|---|---|---|---|---|
| | **Case 1** | | **Case 2** | | **Case 3** | | **Case 4** | |
| | 1P OLTCST | 3P OLTCST | 1P OLTCST | 3P OLTCST | 1P OLTCST | 3P OLTCST | 1P OLTCST | 3P OLTCST |
| $r_a$ | 0.94 | | 0.94 | | 0.94 | | 0.94 | |
| $r_b$ | 0.96 | 0.96 | 0.94 | 0.94 | 0.96 | 0.96 | 0.96 | 0.96 |
| $r_c$ | 0.92 | | 0.92 | | 0.92 | | 0.92 | |
| $P_{loss}\,(W)$ | 19,554.548 | 21,070.299 | 18,607.972 | 19,172.366 | 19,494.120 | 21,003.3412 | 19,201.089 | 20,673.079 |
| Losses (%) | 85.07 | 91.66 | 85.23 | 87.81 | 85.09 | 91.67 | 85.17 | 91.70 |
| $V_{max}\,(V)$ | 246.386 | 243.764 | 246.335 | 245.358 | 246.387 | 243.552 | 246.376 | 243.013 |
| $V_{min}\,(V)$ | 223.084 | 211.088 | 222.224 | 216.103 | 223.182 | 211.199 | 222.896 | 210.921 |
| | **3P-OLTCST** | | | | | | | |
| | **Case 1** | | **Case 2** | | **Case 3** | | **Case 4** | |
| $r_a$ | | | | | | | | |
| $r_b$ | 1 | | 1 | | 1 | | 1 | |
| $r_c$ | | | | | | | | |
| $P_{loss}\,(W)$ | 22,985.288 | | 21,831.880 | | 22,910.126 | | 22,544.249 | |
| Losses (%) | 100 | | 100 | | 100 | | 100 | |
| $V_{max}\,(V)$ | 234.531 | | 230.955 | | 234.310 | | 233.746 | |
| $V_{min}\,(V)$ | 200.288 | | 199.543 | | 200.406 | | 200.106 | |

As in the previous subsection, all these observations strengthen the confirmation of the benefits gained from using decoupled controls and the importance of being rigorous in electrical modeling. In this case, with the earthing configuration, the knowledge of the neutral earthing is needed not only to understand the real state of the system but also to properly solve the voltage optimization problem.

Additionally, in Figure 10, the differences between the bus voltages in Cases 2, 3, and 4 with respect to those of Case 1 are depicted using the 1P-OLTCST. Some noteworthy remarks are as follows: (1) phase-neutral voltages in Case 3 are practically equal to those obtained in Case 1: that is, a ground resistor of 40 Ω produces practically the same voltages as those obtained with a ground resistor of 3 Ω; (2) larger differences occur when a perfectly rigid ground resistance (Case 2) exists, with phase-neutral voltages not as low as those in the case of a non-null ground resistor (Case 1); (3) the final phase-neutral voltages in Case 4 are approximately between the two previous ones.

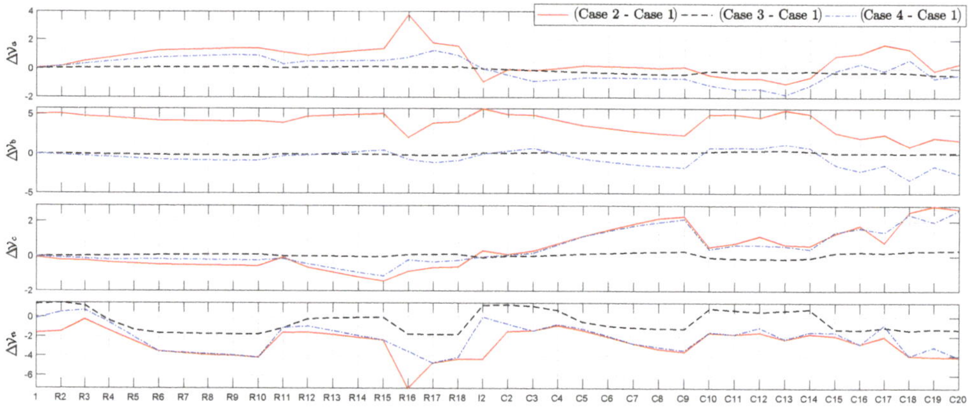


**Figure 10.** Differences between the phase-neutral and neutral bus voltages in Cases 2, 3, and 4 with respect to Case 1.

### 4.4. Distributed Generation

In the previous sections, some simulations are presented in order to demonstrate the importance of the model when carrying out effective control of the power system and for depicting a realistic load scenario without the presence of renewable generation. If distributed generation is considered, the degree of imbalances is expected to be higher, and the differences shown when considering an accurate or a simplified model of the power lines, together with the improvements introduced by the use of a 1P-OLTCST, can become even greater.

In this subsection, distributed PV generation units that can act as single-phase or three-phase generators with regard to their power capability are installed in some buses of the network. A 24-hour time period, during which load and generation profiles are collected every 15 minutes, is considered in this subsection using profiles obtained from [49]. These profiles are adapted to our problem by scaling them to the maximum power and the power factor detailed in Table 7. The distribution of load/generation at each of the three phases of every bus is also specified.

Figure 11 shows the total net apparent power at each hour for the described daily load/generation scenario, so the reader is able to estimate the resulting total imbalances at each time. As expected, the largest imbalances among the three phases are identified as occurring during the hours of highest photovoltaic generation.

The final adopted ground configuration is specified in Table 3.

**Table 7.** Load/distributed generation configuration.

| Bus | R2 | R11 | R15 | R16 | R17 | R18 | I2 | C2 |
|---|---|---|---|---|---|---|---|---|
| $S_{\max}$ $[kVA]$ | 20 | −75 | 25 | −135 | 17 | 80 | −105 | −36 |
| cos $\varphi$ | 0.95 | 0.95 | 0.95 | 0.95 | 0.95 | 0.95 | 0.95 | 0.95 |
| Ph. a (%) | 0 | 30 | 0 | 30 | 100 | 33.33 | 30 | 30 |
| Ph. b (%) | 100 | 40 | 0 | 40 | 0 | 33.33 | 40 | 40 |
| Ph. c (%) | 0 | 30 | 100 | 30 | 0 | 33.33 | 30 | 30 |
| **Bus** | **C12** | **C13** | **C14** | **C17** | **C18** | **C19** | **C20** | |
| $S_{\max}$ $[kVA]$ | −101 | −36 | 35 | 20 | −24 | −48 | −24 | |
| cos $\varphi$ | 0.95 | 0.95 | 0.95 | 0.95 | 0.95 | 0.95 | 0.95 | |
| Ph. a (%) | 30 | 40 | 33.33 | 100 | 30 | 30 | 30 | |
| Ph. b (%) | 40 | 30 | 33.33 | 0 | 40 | 40 | 40 | |
| Ph. c (%) | 30 | 30 | 33.33 | 0 | 30 | 30 | 30 | |

Figure 12 details the obtained optimal tap values when using a 1P-OLTCST or 3P-OLTCST for the specified daily scenario, while Figure 13 shows the evolution of each phase-neutral voltage on the secondary side of the MV/LV transformer. It is easy to identify the numerous times at which the 1P-OLTCST obtains different tap positions for each of the three phases, and it is particularly outstanding for phase b.

In reference to the savings in energy losses, that is, the integral of the objective function reached by the two control strategies, the 1P-OLTCTS results in 3% lower daily losses, while the 3P-OLTCTS saves only 1.05% at the end of the day.

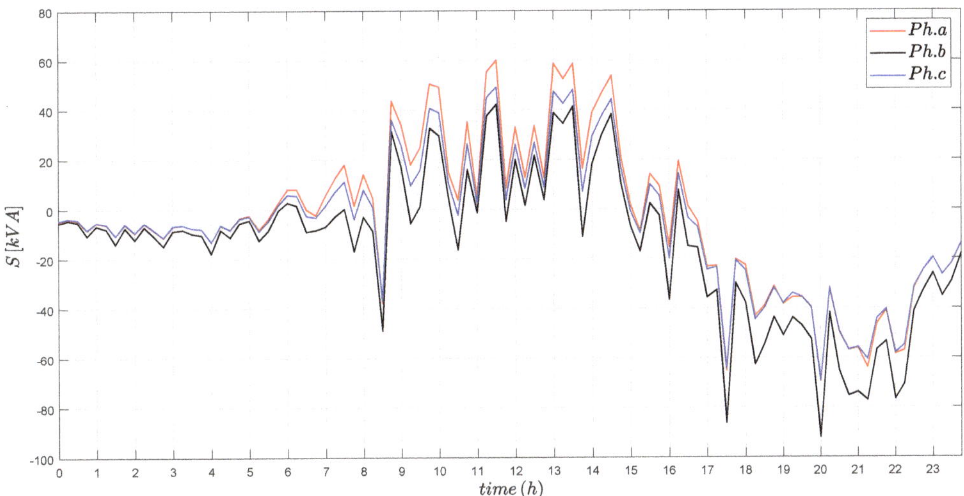


**Figure 11.** Net total apparent power at the MV/LV transformer without considering losses.

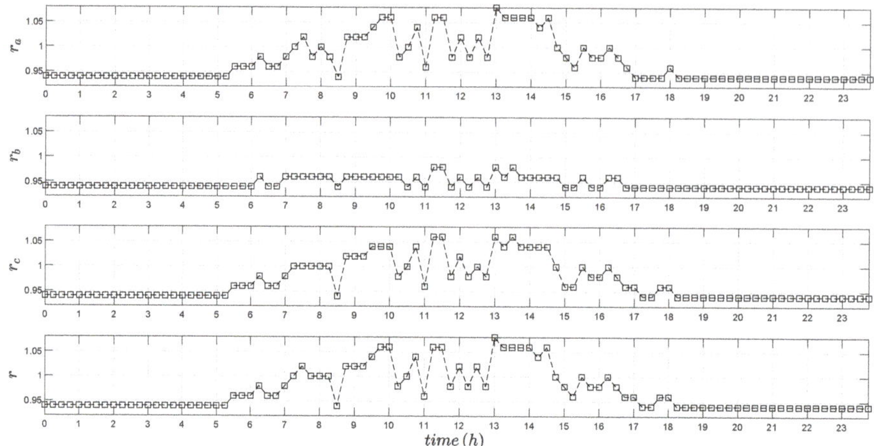

**Figure 12.** Optimal tap positions when considering a 1P-OLTCTS or 3P-OLTCTS.

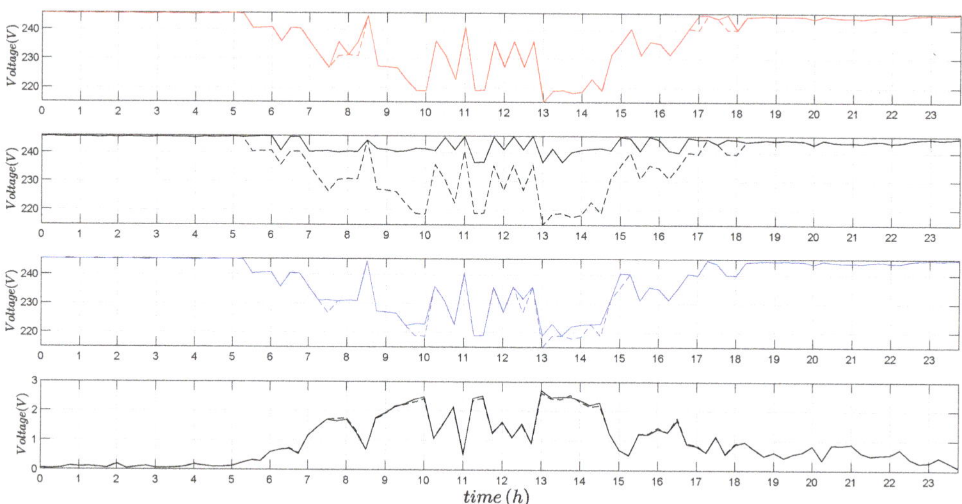

**Figure 13.** Phase-neutral voltage evolution for bus R1 when considering a 1P-OLTCTS or 3P-OLTCTS (dashed lines) for phases a, b, and c and neutral voltage.

Figure 14 shows the maximum and minimum phase-neutral voltage recorded throughout the day at every bus. It is noticeable that the 1P-OLTCTS has lower min–max voltage differences than those obtained with the 3P-OLTCTS. This means that daily voltage variations at any bus are narrower when decoupled voltage control is considered. It is also apparent that the 3P-OLTCTS is not able to maintain voltages within limits, as it exceeds the lower limits at almost all buses at least once a day.

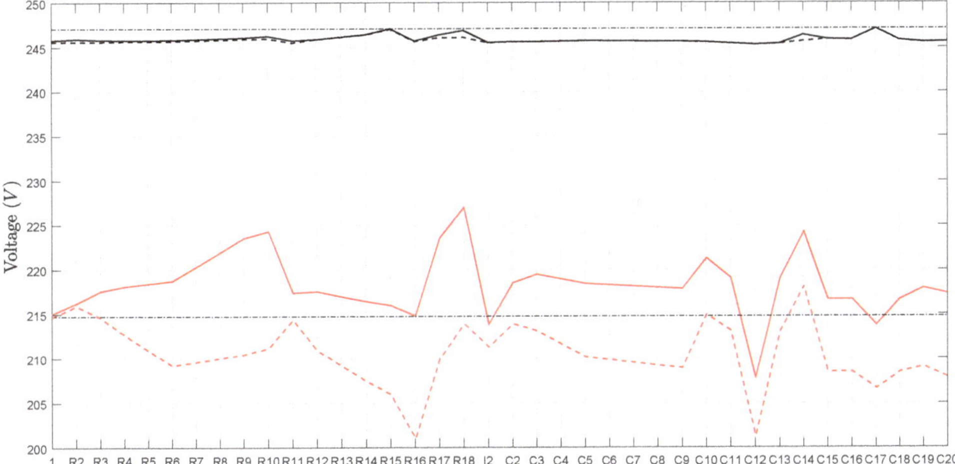


**Figure 14.** Maximum and minimum phase-neutral voltage at every bus for both the 1P-OLTCTS and 3P-OLTCTS (dashed lines).

## 5. Conclusions and Future Work

This work aimed to study optimal voltage control in low-voltage systems by using transformers with on-load tap changers for two options: the classical OLTC that modifies the transformer ratio and a decoupled OLTC that allows different tap positions at each phase of the three-phase transformer. In contrast with previous research, the associated exact optimization problem was detailed and programmed with a special emphasis on the three-phase modeling of all network components as a requirement due to the unbalanced nature of these systems. The proposed tool was tested on a benchmark low-voltage network under different load and renewable generation scenarios. The main conclusions are as follows:

- Using simplified models for three-phase four-wire lines or excluding considerations of the neutral earthing configuration led to erroneous or suboptimal solutions, regardless of whether 1P-OLTCST or 3P-OLTCST was used. This conclusion was obtained even for low unbalanced scenarios (32.1%, 22.7%, and 44.2%), so we can infer that even worse results would occur with an increase in the level of renewable penetration. This conclusion is quite relevant: simplified models are attractive because of their simplicity, but their use can lead to insecure operation of the system.
- The use of simplified models for three-phase four-wires lines caused an underestimation of the voltage drops in most cases, for either of the two OLTCs considered. The worst identified underestimation reached 10 V.
- Regarding the grounding resistance configuration, two main solutions were obtained. It was observed that when the ground resistance increases above zero, the same optimal tap positions are obtained no matter the resistance value (ground resistances of 3 Ω and 40 Ω produce almost the same results, but quite different from those corresponding to $R_g = 0\ \Omega$). Conversely, a different solution is reached when rigid ground connection is taken into consideration.
- Decoupled OLTCs are more effective voltage controls than the common OLTC transformers that change the three-phase transformer ratio. It was proven that the 1P-OLTCTS fulfills its primary function of maintaining the voltage within limits most of the time, even in very unbalanced scenarios. This ability is crucial when analyzing networks with high levels of penetration of renewable energy sources. In contrast, the 3P-OLTCTS did not succeed in fulfilling that requirement in many cases. The definition and implementation of the exact optimization problem

carried out throughout this work was essential to come to this conclusion and would have not been possible if approximated models had been assumed.

- Besides more effectively maintaining voltage within limits and having lower power losses, the 1P-OLTCTS had lower voltage drops and daily voltage variation than the 3P-OLTCTS. For the tested cases, the best saving in power losses was 85.1% for the 1P-OLTCTS, versus 91.7% for the 3P-OLTCTS; the highest voltage drop was 4.5% for the 1P-OLTCTS, versus 9% for the 3P-OLTCTS; and the maximum daily voltage variation was reduced by up to more than 10 V with 1P-OLTCTS versus 3P-OLTCTS.

Future work will examine the coupling of all time frameworks under study in order to reduce the number of movements of tap positions and to maximize the lifetime of the device; it is clear that such tap movements do not lead to profitable savings of power losses and although voltages are within limits, tap movements should be avoided. In addition, the current work focused on the decoupled OLTC as the only means of voltage control in order to understand its real possibilities and because it is a solution that is close to the current reality of utilities. However, additional voltage controls, such as statcoms, inverters linked to renewable generators, or storage systems, would be interesting to study in future analyses.

**Author Contributions:** Conceptualization, Á.R.d.N., E.R.-R., and Á.L.T.-G.; Formal analysis, Á.R.d.N., E.R.-R., and Á.L.T.-G.; Validation, Á.R.d.N., E.R.-R., and Á.L.T.-G.; Writing—original draft, Á.R.d.N., E.R.-R., and Á.L.T.-G.

**Funding:** The authors would like to acknowledge the financial support of the Spanish Ministry of Economy and Competitiveness under grants ENE2014-54115-R and ENE2017-84813-R and the CDTI Grant PASTORA-ITC-20181102 (Innterconecta program, partially FEDER Technology Fund).

**Conflicts of Interest:** The authors declare no conflict of interest.

## References

1. *Country Fiches for Electricity Smart Metering Accompanying the Document Report from the Commission Benchmarking Smart Metering Deployment in the EU-27 with a Focus on Electricity*; Commission Staff Working Document; Publications Office of the European Union: Brussels, Belgium, 2014.
2. International Energy Agency (IEA). *IRENA Cost and Competitiveness Indicators: Rooftop solar PV*; IEA Publications: Paris, France, December 2017.
3. International Energy Agency (IEA). *Global EV Outlook 2017. Two Milllion and Counting*; IEA Publications: Paris, France, 2017.
4. Gozel, T.; Ochoa, L.F. Deliverable 3.3 (Updated) Performance evaluation of the monitored LV networks. In *Electricity North West Limited (ENWL)—Low Voltage Network Solutions*; The University of Manchester: Manchester, UK, July 2014.
5. Mohammadi, P.; Mehraeen, S. Challenges of PV Integration in Low-Voltage Secondary Networks. *IEEE Trans. Power Deliv.* **2017**, *32*, 525–535.
6. Noske, S.; Falkowski, D.; Swat, K.; Boboli, T. UPGRID project: The management and control of LV network. *CIRED-Open Access Proc. J.* **2017**, *2017*, 1520–1522. [CrossRef]
7. Poursharif, G.; Brint, A.; Black, M.; Marshall, M. Analysing the ability of smart meter data to provide accurate information to the UK DNOs. *CIRED-Open Access Proc. J.* **2017**, *2017*, 2078–2081.
8. Wong, P.K.C.; Barr, R.; Kalam, A. Analysis of voltage quality data from smart meters. In Proceedings of the 22nd Australasian Universities Power Engineering Conference (AUPEC), Bali, Indonesia, 26–29 September 2012.
9. Wong, P.K.C.; Barr, R.; Kalam, A. A big data challenge—Turning smart meter voltage quality data into aionable information. In Proceedings of the 22nd International Conference on Electricity Distribution (CIRED), Stockholm, Sweden, 10–13 June 2013.
10. Prado, J.G.; González, A.; Riaño, S. Adopting smart meter events as key data for low-voltage network operation. *CIRED-Open Access Proc. J.* **2017**, *2017*, 924–928. [CrossRef]

11. Navarro, A.; Ochoa, L.F.; Mancarella, P.; Randles, D. Impacts of photovoltaics on low voltage networks: A case study for the North West of England. In Proceedings of the 22nd International Conference on Electricity Distribution (CIRED), Stockholm, Sweden, 10–13 June 2013.
12. Report C1: Use of smart meter information for network planning and operation. In *UK Power Networks Holdings Limited—Low Carbon London Project*; UK Power Networks Holdings Limited: London, UK, September 2014.
13. Seymour, J. The Seven Types of Power Problems. White Paper 18, Revision 1, Schneider Electric—Data Center, Science Center. 2011. Available online: http://www.apc.com/salestools/VAVR-5WKLPK/VAVR-5WKLPK_R1_EN.pdf (accessed on 1 May 2019).
14. O'Connell, A.; Soroudi, A.; Keane, A. Distribution Network Operation Under Uncertainty Using Information Gap Decision Theory. *IEEE Trans. Smart Grid* **2018**, *9*, 1848–1858.
15. Lan, B.-R.; Chang, C.-A.; Huang, P.-Y.; Kuo, C.-H.; Ye, Z.-J.; Shen, B.-C.; Chen, B.-K. Conservation voltage regulation (CVR) applied to energy savings by voltage-adjusting equipment through AMI. *IOP Conf. Ser. Earth Environ. Sci.* **2017**, *93*, 012070. [CrossRef]
16. Nijhuis, M.; Gibescu, M.; Cobben, J.F.G. Valuation of measurement data for low voltage network expansion planning. *Electr. Power Syst. Res.* **2017**, *151*, 59–67. [CrossRef]
17. Aziz, T.; Ketjoy, N. PV Penetration Limits in Low Voltage Networks and Voltage Variations. *IEEE Access* **2017**, *5*, 16784–16792.
18. Tahir, M.; Nassar, M.E.; El-Shatshat, R.; Salama, M.M.A. A review of Volt/Var control techniques in passive and active power distribution networks. In Proceedings of the 4th IEEE International Conference on Smart Energy Grid Engineering (SEGE), Oshawa, ON, Canada, 21–24 August 2016; pp. 57–63.
19. Bayer, B.; Matschoss, P.; Thomas, H.; Marian, A. The German experience with integrating photovoltaic systems into the low-voltage grids. *Renew. Energy* **2018**, *119*, 129–141. [CrossRef]
20. Yan, R.; Marais, B.; Saha, T.K. Impacts of residential photovoltaic power fluctuation on on-load tap changer operation and a solution using DSTATCOM. *Electr. Power Syst. Res.* **2014**, *111*, 185–193. [CrossRef]
21. Long, C.; Ochoa, L.F. Voltage Control of PV-Rich LV Networks: OLTC-Fitted Transformer and Capacitor Banks. *IEEE Trans. Power Syst.* **2016**, *31*, 4016–4025. [CrossRef]
22. Coppo, M.; Turri, R.; Marinelli, M.; Han, X. Voltage management in unbalanced low voltage networks using a decoupled phase-tap-changer transformer. In Proceedings of the 49th International Universities Power Engineering Conference (UPEC), Cluj-Napoca, Romania, 2–5 September 2014; pp. 1–6.
23. Zecchino, A.; Marinelli, M.; Hu, J.; Coppo, M.; Turri, R. Voltage control for unbalanced low voltage grids using a decoupled-phase on-load tap-changer transformer and photovoltaic inverters. In Proceedings of the 50th International Universities Power Engineering Conference (UPEC), Stoke on Trent, UK, 1–4 September 2015; pp. 1–6.
24. Hu, J.; Marinelli, M.; Coppo, M.; Zecchino, A.; Bindner, H.W. Coordinated voltage control of a decoupled three-phase on-load tap changer transformer and photovoltaic inverters for managing unbalanced networks. *Electr. Power Syst. Res.* **2016**, *131*, 264–274. [CrossRef]
25. Zecchino, A.; Hu, J.; Coppo, M.; Marinelli, M. Experimental testing and model validation of a decoupled-phase on-load tap-changer transformer in an active network. *IET Gen. Transm. Distrib.* **2016**, *10*, 3834–3843. [CrossRef]
26. Quirós-Tortós, J.; Ochoa, L.F.; Alnaser, S.W.; Butler, T. Control of EV Charging Points for Thermal and Voltage Management of LV Networks. *IEEE Trans. Power Syst.* **2016**, *31*, 3028–3039. [CrossRef]
27. Efkarpidis, N.; Rybel, T.D.; Driesen, J. Optimal Placement and Sizing of Active In-Line Voltage Regulators in Flemish LV Distribution Grids. *IEEE Trans. Ind. Appl.* **2016**, *52*, 4577–4584. [CrossRef]
28. Procopiou, A.T.; Ochoa, L.F. Voltage Control in PV-Rich LV Networks Without Remote Monitoring. *IEEE Trans. Power Syst.* **2017**, *32*, 1224–1236. [CrossRef]
29. Pezeshki, H.; Arefi, A.; Ledwich, G.; Wolfs, P. Probabilistic Voltage Management Using OLTC and dSTATCOM in Distribution Networks. *IEEE Trans. Power Deliv.* **2017**, *33*, 570–580. [CrossRef]
30. Fallahzadeh-Abarghouei, H.; Nayeripour, M.; Hasanvand, S.; Waffenschmidt, E. Online hierarchical and distributed method for voltage control in distribution smart grids. *IET Gen. Transm. Distrib.* **2017**, 11, 1223–1232 [CrossRef]
31. Aziz, T.; Ketjoy, N. Enhancing PV Penetration in LV Networks Using Reactive Power Control and On Load Tap Changer with Existing Transformers. *IIEEE Access* **2018**, *6*, 2683–2691. [CrossRef]

32. Bettanin, A.; Coppo, M.; Savio, A.; Turri, R. Voltage management strategies for low voltage networks supplied through phase-decoupled on-load-tap-changer transformers. In Proceedings of the AEIT International Annual Conference, Cagliari, Italy, 20–22 September 2017; pp. 1–6.
33. Jiricka, J.; Kaspirek, M.; Kolar, L.; Zahradka, M. Smart Substation MV/LV. In Proceedings of the 24th International Conference on Electricity Distribution (CIRED), Glasgow, UK, 12–15 June 2017; pp. 1–5.
34. Baudot, C.; Roupioz, G.; Carre, O.; Wild, J.; Potet, C. Experimentation of voltage regulation infrastructure on LV network using an OLTC with a PLC communication system. *CIRED-Open Access Proc. J.* **2017**, *2017*, 1120–1122. [CrossRef]
35. Mokkapaty, S.; Weiss, J.; Schalow, F.; Declercq, J. New generation voltage regulation distribution transformer with an on load tap changer for power quality improvement in the electrical distribution systems. *CIRED-Open Access Proc. J.* **2017**, *2017*, 784–787. [CrossRef]
36. Daratha, N.; Das, B.; Sharma, J. Coordination Between OLTC and SVC for Voltage Regulation in Unbalanced Distribution System Distributed Generation. *IEEE Trans. Power Syst.* **2014**, *29*, 289–299. [CrossRef]
37. Frame, D.; Bell, K.; McArthur, S. A Review and Synthesis of the Outcomes from Low Carbon Networks Fund Projects. UKERK Publications. August 2016. Available online: http://www.ukerc.ac.uk/publications/a-review-and-synthesis-of-the-outcomes-from-low-carbon-networks-fund-projects.html (accessed on 1 May 2019).
38. Chen, T.-H.; Yang, W.-C. Analysis of multi-grounded four-wire distribution systems considering the neutral grounding. *IEEE Trans. Power Deliv.* **2001**, *16*, 710–717. [CrossRef]
39. Urquhart, A.J. Accuracy of Low Voltage Electricity Distribution Network Modelling. Ph.D. Thesis, Loughborough Universtiy, Loughborough, UK, 2016.
40. Willis, H. L. *Power Distribution Planning Reference Book*, 2nd ed.; CRC Press: Boca Raton, FL, USA, March 2004; ISBN 978-0824748753.
41. Özkan, Z.; Hava, A.M. Three-phase inverter topologies for grid-connected photovoltaic systems. In Proceedings of the International Power Electronics Conference (IPEC), Hiroshima, Japan, 18–21 May 2014; pp. 498–505.
42. Moreno-Díaz, L.; Romero-Ramos, E.; Gómez-Expósito, A.; Cordero-Herrera, E.; Rivero, J.R.; Cifuentes, J.S. Accuracy of Electrical Feeder Models for Distribution Systems Analysis. In Proceedings of the International Conference on Smart Energy Systems and Technologies (SEST), Sevilla, Spain, 10–12 September 2018; pp. 1–6.
43. Chen, M.S.; Dillon, W.E. Power system modeling. *Proc. IEEE* **1974**, *62*, 901–915. [CrossRef]
44. Rodas, R.D.E.; Padilha-Feltrin, A.; Ochoa, L.F. Distribution transformers modeling with angular displacement: Actual values and per unit analysis. *SBA Controle & Automação Sociedade Brasileira de Automatica* **2007**, *18*, 499–500.
45. Carneiro, S.; Martins, H.J.A. Measurements and model validation on three-phase core-type distribution transformers. In Proceedings of the IEEE Power Engineering Society General Meeting, Toronto, ON, Canada, 13–17 March 2004; pp. 120–124.
46. Kersting, W.H. *Distribution System Modeling and Analysis*, 3rd ed.; CRC Press: Boca Raton, FL, USA, February 2016.
47. *Modelling and Aggregation of Loads in Flexible Power Networks*; Working Group C4.605; CIGRÉ, Conseil International des Grands Réseaux éLectriques: Paris, France, February 2014.
48. *Benchmark Systems for Network Integration of Renewable and Distributed Energy Resources*; Comité d'études C6; CIGRÉ, Conseil international des grands réseaux électriques: Paris, France, 2014.
49. Gozel, T.; Navarro, A.; Ochoa, L.F. Deliverable 3.5 Creation of Aggregated Profiles with and without New Loads and DER Based on Monitored Data. April 2014. Available online: https://www.enwl.co.uk/zero-Carbon/smaller (accessed on 1 May 2019).

*Article*

# BPL-PLC Voice Communication System for the Oil and Mining Industry

**Grzegorz Debita [1,*], Przemysław Falkowski-Gilski [2,*], Marcin Habrych [3], Grzegorz Wiśniewski [3], Bogdan Miedziński [3], Przemysław Jedlikowski [4], Agnieszka Waniewska [1], Jan Wandzio [5] and Bartosz Polnik [6]**

1 Faculty of Economics, General Tadeusz Kosciuszko Military University of Land Forces, Czajkowskiego St. 109, 51-147 Wroclaw, Poland; agnieszka.waniewska@awl.edu.pl
2 Faculty of Electronics, Telecommunications and Informatics, Gdansk University of Technology, Narutowicza St. 11/12, 80-233 Gdansk, Poland
3 Faculty of Electrical Engineering, Wroclaw University of Science and Technology, Wybrzeze Wyspianskiego St. 27, 50-370 Wroclaw, Poland; marcin.habrych@pwr.edu.pl (M.H.); grzegorz.wisniewski@pwr.edu.pl (G.W.); bogdan.miedzinski@pwr.edu.pl (B.M.)
4 Faculty of Electronics, Wroclaw University of Science and Technology, Wybrzeze Wyspianskiego St. 27, 50-370 Wroclaw, Poland; przemyslaw.jedlikowski@pwr.edu.pl
5 KGHM Polska Miedz S.A. Sklodowskiej-Curie St. 48, 59-301 Lubin, Poland; jan.wandzio@kghm.com
6 KOMAG Institute of Mining Technology, Pszczynska St. 37, 44-101 Gliwice, Poland; bpolnik@komag.eu
* Correspondence: grzegorz.debita@awl.edu.pl (G.D.); przemyslaw.falkowski@eti.pg.edu.pl (P.F.-G.)

Received: 10 August 2020; Accepted: 8 September 2020; Published: 12 September 2020

**Abstract:** Application of a high-efficiency voice communication systems based on broadband over power line-power line communication (BPL-PLC) technology in medium voltage networks, including hazardous areas (like the oil and mining industry), as a redundant mean of wired communication (apart from traditional fiber optics and electrical wires) can be beneficial. Due to the possibility of utilizing existing electrical infrastructure, it can significantly reduce deployment costs. Additionally, it can be applied under difficult conditions, thanks to battery-powered devices. During an emergency situation (e.g., after coal dust explosion), the medium voltage cables are resistant to mechanical damage, providing a potentially life-saving communication link between the supervisor, rescue team, paramedics, and the trapped personnel. The assessment of such a system requires a comprehensive and accurate examination, including a number of factors. Therefore, various models were tested, considering: different transmission paths and types of coupling (inductive and capacitive), as well as various lengths of transmitted data packets. Next, a subjective quality evaluation study was carried out, considering speech signals from a number of languages (English, German, and Polish). Based on the obtained results, including both simulations and measurements, appropriate practical conclusions were formulated. Results confirmed the applicability of BPL-PLC technology as an efficient voice communication system for the oil and mining industry.

**Keywords:** audio coding; digital systems; electrical engineering; ICT; Industry 4.0; IoT; power cable; QoS; reliability; voice communication

---

## 1. Introduction

In recent years, the use of power line communication (PLC) technology has become increasingly common. Although PLC technology was introduced many years ago, it is still being developed and successfully used in various areas, including both low and medium voltage scenarios. An example of this may be its wide use in advanced metering infrastructure (AMI) systems for various types of electrical measurements, as well as for remote control of electricity consumption by end-users [1–6]. This is primarily due to the possibility of utilizing an already existing infrastructure, including both low as well as high voltage power grids. As a result, the investment costs of the installation are significantly reduced and operator fees can be omitted. However, the disadvantages of PLC transmission include, most of all, the impact of environmental electromagnetic compatibility (EMC) disturbances, both conducted and induced, on the scope and quality of transmission [7–11]. In addition, PLC transmission should not interfere with the basic functions of electricity networks, such as transmission and distribution of electricity itself. It should therefore adapt to the current technical conditions and operation restrictions of these networks.

The use of the broadband over power line (BPL) technology is obviously limited, due to possible interference with other systems. BPL solutions available on the market enable transmission only for strictly defined frequency bands. It should be also emphasized that different telecommunication operators, especially those related to wireless transmission systems (e.g., Wi-Fi and various terrestrial radio communication standards), are not particularly interested in the development of PLC technology.

Authors, familiar with the problem of maintaining communication in harsh environments, especially underground mines, have found that the use of BPL technology, particularly in MV cable networks, can be beneficial. BPL enables signal transmission within a frequency range of 2–30 MHz, with bitrates even up to 200 Mbps. This is much faster than narrowband communication. The BPL-PLC technology can be successfully used both for control and monitoring of environmental conditions, and secondary as voice and/or image transmission of the so-called last chance, especially during mine disaster or other random events [9–11]. The underground cable lines are above all the most resistant to mechanical damage and thus can provide an effective transmission channel under various hazardous situations. Of course, in this case, battery-powered devices are mandatory.

However, to provide an effective BPL transmission in a particular network, it is necessary to examine and explain the impact of a number of factors, both external (harsh environment) and internal, related to the physical parameters of all network components (cables, permanent joints, connectors, bus bars, etc.), structure and network loading conditions (value and nature of distorted electrical quantities). It should be noted that the influence of the above-mentioned factors is variable over time and can be unpredictable [12–15].

The disadvantages of PLC transmission in low-voltage networks are recognized, widely analyzed, and well-known in literature [7,16]. In medium voltage networks, particularly in mines, the main problem in performing field tests is limited access, related to electrical safety. Therefore, it may be sometimes difficult to derive the physical quantities of the analyzed cable. Due to this, the verification of the propagation model becomes a major issue to overcome [9–11]. Based on previous research conducted by the authors, it has been shown that the efficiency of high-frequency transmission in underground cable mine networks can be subjected to significant changes, depending on the network load conditions (with or without supply, under-voltage but unloaded, loaded, etc.), and this is not necessarily caused by noise, as the bridged effect of energy consumers can be neglected [17–19].

It is particularly important to select effective ways of coupling modems with the cable line, taking into account the frequency of transmitted signals and the need to match to the characteristic impedance of the wired medium. However, one should be aware that the way of coupling modems with a power cable affects the signal path over the cable, and thus the quality of transmission. Therefore, when analyzing the effectiveness of the BPL-PLC technology, an appropriate contribution of the cable's electrical parameters (resistance, capacity, inductance, conductivity) should be taken into account, depending on the type of coupling (the term coupling is used here to match or mismatch with a suitable coupler device [16]). In practice, this task is not a simple matter. This applies especially to mixed-coupling, e.g., capacitive-inductive. Further improvements regarding coupling solutions are discussed in [20].

The main purpose of carrying out research was the need to find redundant means of communication, compared to a traditional fiber optic or cooper media, which would prove to be not only stable but also resistant to mechanical damage, i.e., the impact of rock mass. During an emergency situation, medium voltage cables are hard to take damage. This makes them an ideal solution for maintaining a high-efficiency voice communication between miners and/or paramedic team during any rescue operation. It should be noted that the underground electrical network is different from the electrical networks used on the surface. The tested mining cable itself was reinforced with a galvanized steel tape and had a general screen.

The impact of the cable's electrical parameters on the transmission quality for both inductive-inductive and capacitive-inductive types of coupling was analyzed and confirmed by field tests, which is a novelty. It should be emphasized that the mathematical models were appropriately adjusted by the authors (Section 3). Additionally, the results of the research presented in Section 4.1 were obtained using our custom-developed software. The results of theoretical research have been verified by tests under real-time operating conditions in a genuine mine, which is a significant added value. The transmission efficiency, regarding the amount of retransmitted data, depending on the TCP/IP window size, packet size, and the network latency for UDP, was also analyzed. Based on the obtained results, appropriate practical conclusions were made.

This work is an extended version and a direct continuation of our published conference paper [21]. It summarizes many years of research conducted by the authors, hence some of the results were published early [9–11,21]. The content of Section 3, as well as Sections 4.1 and 4.3, has not been previously published. Results describing the quality evaluation of transmitted speech signal samples, including both couplings (inductive-inductive and capacitive-inductive), four speakers (two female and two male individuals), and transmission routes (A to B and vice versa) have been averaged overall, to present a comprehensive subjective analysis of the described communication system. The remaining sections, that is an introduction and description of BPL-PLC technology, have been extended and revised, respectively.

The manuscript consists of 5 sections. Section 1 is an introduction to the discussed research subject along with the justification for the choice of this particular topic. Section 2 presents a fragment of the medium voltage mining network, detailing the cable on which the tests were carried out, including the coupling method. Section 3 analyzes the influence of individual electrical parameters of the cable and their change with frequency for five selected simulation models, regarding the length of the transmitted data packet, as well as a quality evaluation study of voice transmission. The results of analyzes and considerations along with the discussion are presented in Section 4. Section 5 summarizes and concludes this manuscript.

## 2. Materials and Methods

A part of the 6 kV mining network, selected for analysis and testing, along with indicated modem location, is shown in Figure 1. This cable line was a part of an underground power grid with a total length of over 2.5 km, on which BPL-PLC implementation works were carried out. The analyzed mining cable was a YKGYFtznyn 6/3.6 kV 3 × 185 $mm^2$ type model, reinforced with a galvanized steel tape and a general screen. During the investigation, it was assumed that factors such as load or passives had no influence. Due to the type of utilized filters, the study was carried out in the 2–18 MHz frequency range. The utilized modems were bought freely on the market.

The cable under study, shown in Figure 2, was a 3-phase mine cable (YKGYFtZnyn 3 × 185 $mm^2$) with an inner copper screen equipped with a steel armor to protect it. The L1, L2, and L3 phases separated uniformly from each other were made of stranded copper conductors with PVC insulation. The electrical parameters of the cable have been derived based on the knowledge concerning the cable's structure, the cross-sectional geometrical dimensions, and properties of the used materials (conductive or insulating). The tested cable is specific due to its structure. It has been designed and made particularly for this kind of environment (mines). It differs from other common power cables intended for numerous applications, mainly it is a 3-phase model and has a steel armor. The solution of the cable earthing system and the associated signal flow are specific to mine applications.

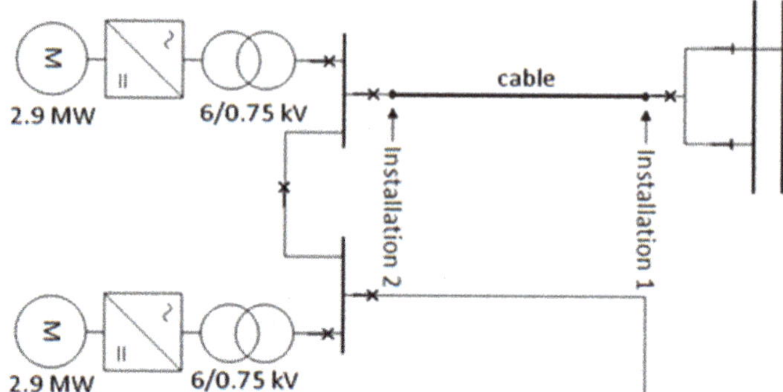


**Figure 1.** Analyzed section of the 6 kV cable network with connected broadband over power line (BPL) devices.

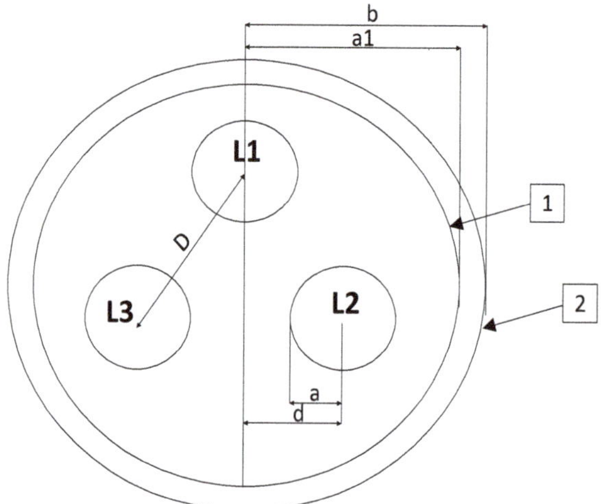


**Figure 2.** Cross-sectional view of the 3-phase 6 kV power cable: 1–screen (made of copper), 2–armor (steel composition), L1, L2, L3–phase conductors (made of copper); a = 15 mm, D = 46.96 mm, b = 63 mm, a1 = 55 mm, d = 23.48 mm; screen material conductivity: $58.67 \times 10^6$ 1/Ωm, armor material conductivity: $10.02 \times 10^6$ 1/Ωm, polyvinyl as an insulating filing medium: $10^{-14}$ 1/Ωm; electric permittivity of polyvinyl: 3.3; magnetic permeability of copper 0.9999, steel composition: 60.

Theoretical considerations and field tests were carried out for both inductive-inductive and mixed capacitive-inductive couplings, as shown in Figure 3. It should be emphasized that in mine electric networks the use of capacitive coupling is often troublesome, due to lack of access to phase conductors. This is particularly related to cables routed in shafts. In addition, capacitive type couplings are not recommended for use in medium voltage cable networks, due to electrical safety requirements in mines. In BPL-PLC solutions, such coupler should be connected directly to the phase/phases, without the use of any other short-circuit protection (fuse). This implies a high risk of both damage to equipment and electric shock to staff.

Therefore, the only way in such cases is to connect the capacitive coupler to the bus/phase, but through a thin enough electric wire of a sufficient length, to meet requirements of the electrical strength of the gap. Otherwise, the wire would simply evaporate under fault current. Additional information on BPL-PLC solutions, particularly related to smart grids, data transfer, as well as service quality estimation, may be found in [22–25].

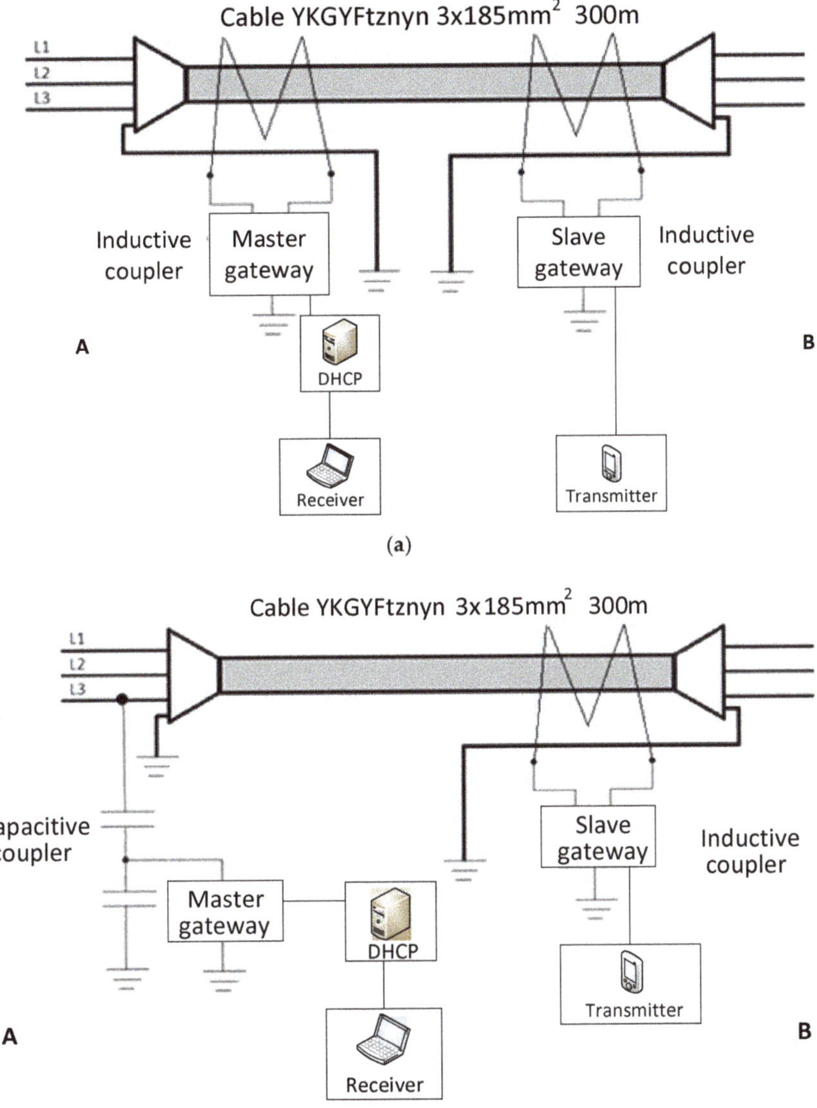


**Figure 3.** *Cont.*

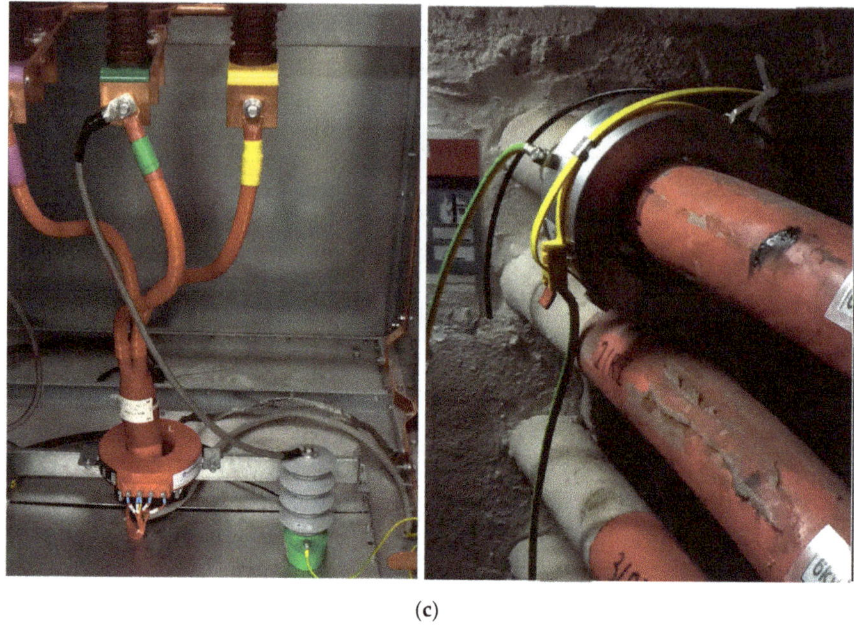

(c)

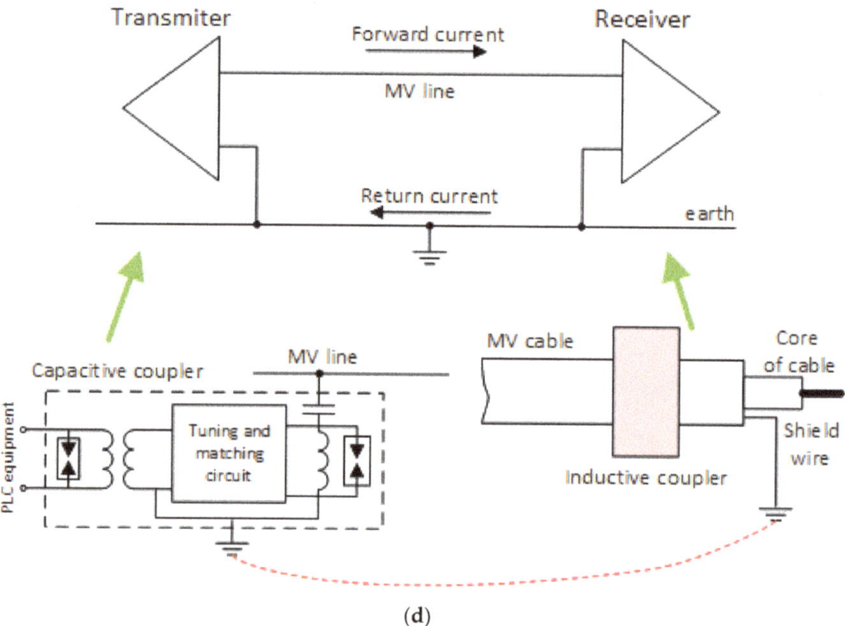


(d)

**Figure 3.** Layout for the BPL transmission tests: (**a**) Inductive-inductive coupling; (**b**) Capacitive-inductive coupling; (**c**) Physical installation of capacitive and inductive couplers; (**d**) Schematic of the signal propagation.

## 3. Methodology

Simulation models described in this section consider the impact of wave propagation and packet length analysis in the wired medium, as well as a subjective quality evaluation study of real-time transmitted speech samples.

The objective parameter, based on which it was possible to practically verify the energy balance of the link, is the attenuation of the transmission medium. Therefore, to estimate the impact of the physical parameters of the cable with frequency, the authors developed the presented models and compared the obtained results with carried out measurements. It should be emphasized that only the model, considering the dependence of longitudinal resistance of the cable with frequency, is practically commonly used (approach no. 1). The basic model, known from the literature [9], has been therefore appropriately modified to qualitatively and quantitatively assess the influence of selected physical parameters of the cable and their variation with frequency on the propagation conditions of the medium voltage 3-phase mining cable. Hence, the presented analysis of different 5 models (approaches).

Five theoretical models were used to consider the impact of the cable's electrical parameters and frequency for the inductive-inductive and mixed capacitive-inductive coupling. In the first model (approach no. 1, propagation constant according to Equation (1)), only the variation of longitudinal cable resistance $R(f)$ with respect to frequency was adopted for analysis. Other values of electrical quantities were calculated from equivalent circuits, taking into account the geometry of the utilized cable.

$$\gamma = \sqrt{(R(f) + j2\pi fL)(G + j2\pi fC)} \tag{1}$$

where: $L$, $G$, $C$–transversal parameters of the cable.

To perform simulations for the second model (approach no. 2, propagation constant according to Equation (2)), the longitudinal cable conductance $G(f) = 1/R(f)$ was assumed to be a function of frequency as well. Therefore, the propagation constant $\gamma$ could be derived from formula [25]:

$$\gamma = \sqrt{(2\pi fj\mu(G(f) + 2\pi fj\mu\varepsilon))}, \tag{2}$$

where: $\mu$–magnetic permeability for copper, $\varepsilon$–dielectric constant for polyvinyl.

The third model (approach no. 3) takes into account the dependency of longitudinal resistance $R(f)$ as well as transversal conductivity $G'(f)$ of the cable with respect to frequency. The propagation constant $\gamma$ was therefore derived using the transmission line theory [1,9,16]:

$$\gamma = \sqrt{(R(f) + j2\pi fL)(G'(f) + j2\pi fC)}. \tag{3}$$

Taking the transverse conductivity value $G'$ for polyvinyl, a large discrepancy with the results of attenuation measurements for the analyzed cable was observed. The highest compliance was found when assuming the value of this conductivity equal to the reciprocal of the longitudinal resistance of the cable, i.e.,

$$G = \frac{1}{R(f)} \sim G'(f). \tag{4}$$

It should be noted that the convergence of the results increases with the length of the cable.

The fourth model (approach no. 4) assumed the dependency of $R$, $G'$ and $C$ on frequency. The propagation constant was calculated as

$$\gamma = \sqrt{(R(f) + j2\pi fL)(G'(f) + j2\pi fC(f))}, \tag{5}$$

where: $C(f)$ was determined analytically based on knowledge of all physical parameters and cable dimensions:

$$C(f) = \frac{\varepsilon G(f)}{\sigma};\ G'(f) = G(f), \tag{6}$$

where: $\varepsilon, \sigma$–respective values for polyvinyl.

In the last model (approach no. 5), all electrical quantities of the cable were assumed to change with respect to frequency. The propagation constant was determined as for transmission line Equation (5). Whereas, the cable's inductance $L(f)$, including skin effect, was calculated from the formula:

$$L(f) = \frac{\varepsilon\mu}{C(f)}, \tag{7}$$

where: $\varepsilon$–value for polyvinyl, whereas $\mu$ for copper, respectively.

The analysis of the efficiency of BPL-PLC transmission also required the respective selection of the IP protocol in the TCP version and connectionless UDP [26,27]. Therefore, we have measured and analyzed the influence of the amount of retransmissions, depending on the TCP window size, packet size, and network latency for UDP. Later on, a subjective study was carried out, considering speech samples sourced from ITU-T P.501 [28]. This dataset included signal samples consisting of two sentences spoken by two female and two male individuals, in different languages. Due to the international character of the oil and mining industry, we have selected samples from 4 sets, namely: American English (AE), British English (EN), German (GE), and Polish (PL).

## 4. Results and Discussion

### 4.1. Wave Propagation

Selected results of analyzes and measurements are shown in Figure 4. According to Figure 4a, the variation of the calculated attenuation coefficient $\alpha$ versus cable length shows that the highest compliance with the measured data is obtained when using the theoretical approach in accordance with models 3–5 (approach no. 3–5) (the area between the red dotted lines represents the theoretical extent of regression error). The measurements were carried out under varying location of the output (inductive) coupler, in relation to the input coupler during different current load of the tested cable. The attenuation was measured using the dedicated software from the modem's provider. As shown, the measurement point for 300 m differs significantly from the others, due to (most probably) the influence of the deformed wave of the load current of the tested cable. It should be noted that measurements were performed during one single day at a stable ambient temperature. However, the temperature of the cable could (of course) undergo some changes as a function of its current load. As a result, this could have some influence on the obtained result.

The calculations deviate the least from the regression curve in case of the considered inductive-inductive coupling. However, the simplest way to estimate the attenuation seems to be by means of model 5 (approach no. 5).

Knowing the structure and geometrical dimensions of the cable, equivalent values of resistance, conductivity, capacity, and inductance, were calculated, respectively. Next, the remaining data was derived using Equations (2)–(6). As for the characteristic impedance $Z_0$ of the cable, the results obtained using approach no. 2 differ from the others, as shown in Figure 4b. The best approximation was also found for calculations based on model 5 (approach no. 5).

Approach no. 2 was adopted only due to the change of conductance of the cable. We have assumed that conductance was a function of frequency. We have also stated that if the conductance was calculated only from the geometric parameters of the cable (without the skin effect), the simulations differed significantly from the measurements. As shown in Figure 4, the result corresponding to approach no. 2 was marked in yellow, whereas the blue color was related to model 1 (approach no. 1), which takes into account only the change in resistance as a function of frequency. The dependencies derived based on model 3 (approach no. 3), taking into account variation of both resistance and conductance with respect to frequency, were marked in red. The green color applied to model 4 (approach no. 4), which takes into account the dependence of capacity on frequency. Model 5 (approach no. 5) takes into account the change in all electrical parameters as a function of frequency.

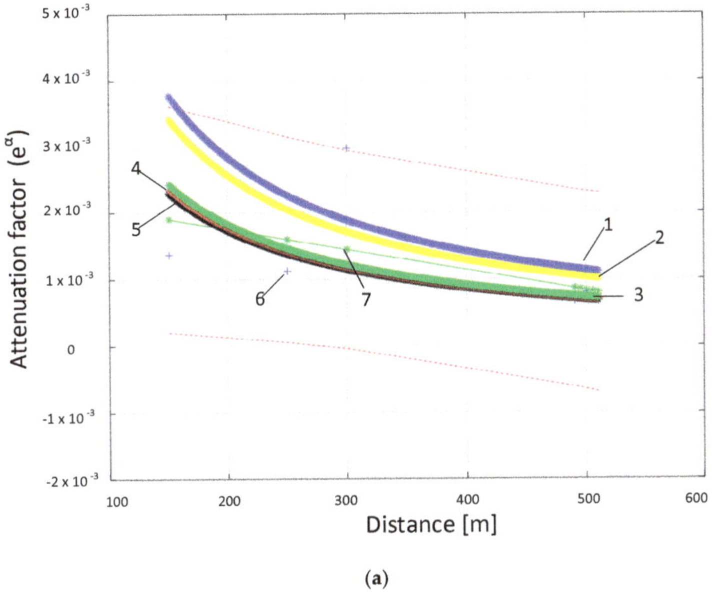


(a)

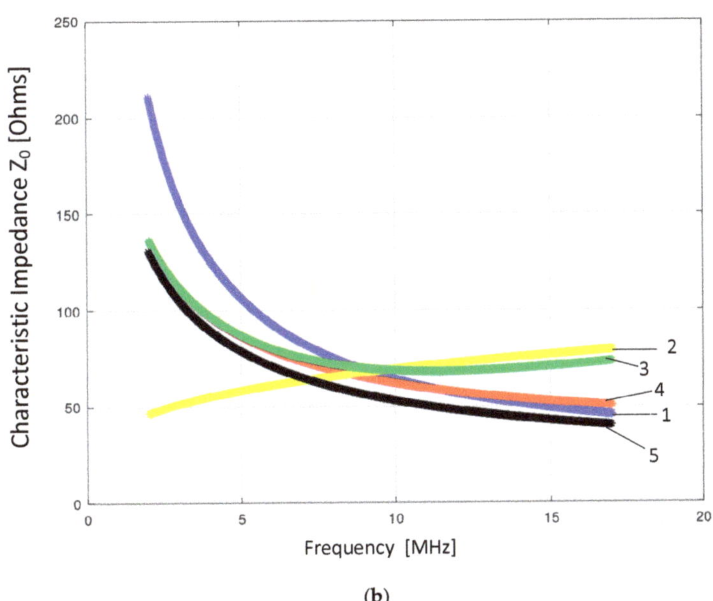


(b)

**Figure 4.** *Cont.*

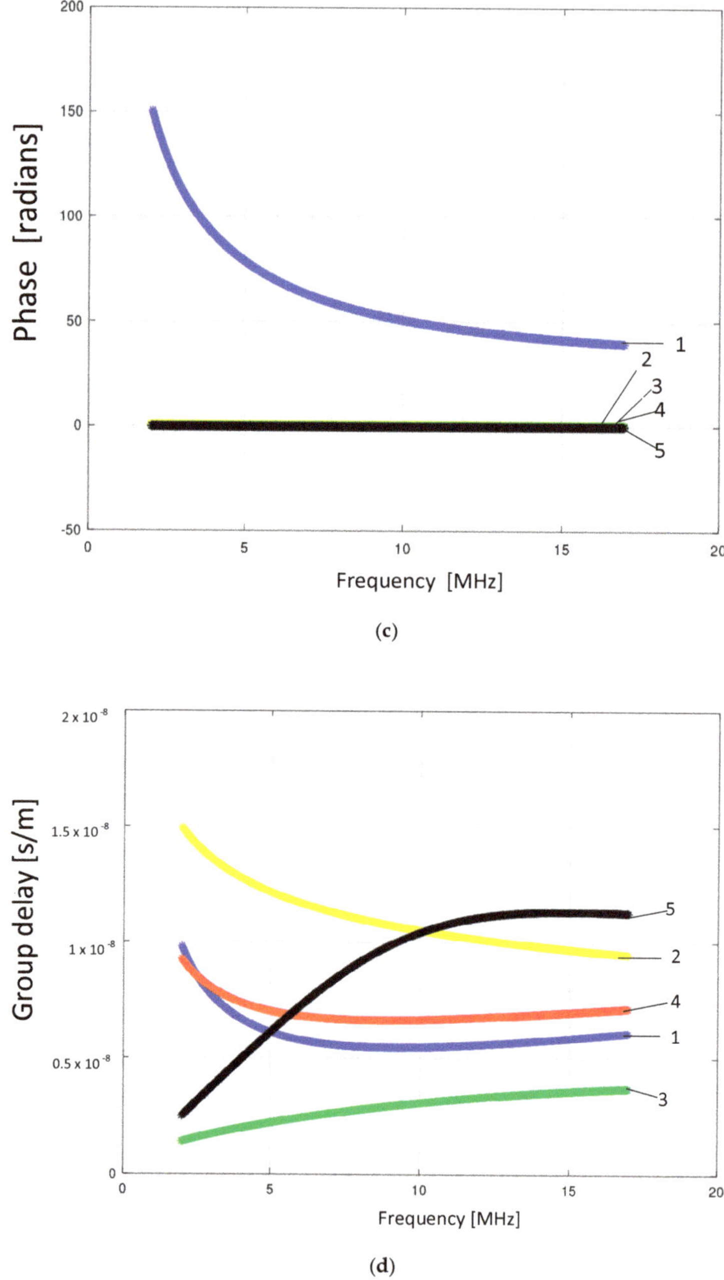


**Figure 4.** The impact of the cable's electrical parameters and frequency for the inductive-inductive coupling: (**a**) Comparison of measured and derived values of attenuation factor $\alpha$ as a function of cable length (1–5—analyzed approaches, 6–measurement results, 7–regression line); (**b**) Calculated characteristic impedance $Z_0$ of the analyzed cable as a function of frequency in 5 analyzed approaches; (**c**) Calculated phase value as a function of frequency in 5 analyzed approaches; (**d**) Calculated group delay versus frequency in 5 analyzed approaches.

It was found that the phase of the signal does not change with frequency and is close to zero, as can be seen from the curves shown in Figure 4c. This only differs for calculations made in accordance with model 1 (approach no. 1). In this model, only the dependency of resistance on frequency was assumed.

According to calculations (based on model no. 5), the group delay tends to increase, as shown in Figure 4d. This is in accordance with results regarding the impact of frame length on transmission efficiency. Therefore, it seems that the analysis of the effectiveness of BPL transmission in a medium voltage power cable, with the inductive-inductive coupling, according to model 5 (approach no. 5), gives the most reliable results.

The analysis of transmission efficiency in a medium voltage mine cable, with the mixed capacitive-inductive coupling, shown in Figure 5, was performed using similar propagation modeling methods (approaches no. 1–5) as for the inductive-inductive ones. However, due to the specificity of this coupling, the contribution of resistance $R$ of only one phase was adopted for analysis (the model is calculated for a 3-phase cable system, but the capacitive coupler is connected to one phase).

The effect of the resistance of other cable elements, such as shielding ($R_s$) and armor ($R_a$), has been neglected. The assumption that $R_s$ and $R_a$ are omitted is related to the fact that signals were launched directly into the selected phase for capacitive coupling. Figure 3c shows that the input capacitive coupler is connected to one phase, whereas the inductive coupler (installed anywhere) covers the entire cable. The closed path for transmitted signals is performed through the phase and the cable earthing system, as shown in Figure 3b. This fact was of course taken into account when modeling the cable. As shown in Figure 3c, there is not enough space inside the switchgear to install an inductive coupler (including, apart from the phase conductors, also the grounding conductor/screen). In this situation, it was necessary to use a capacitive coupler. This coupler must be connected directly to the selected phase without using any fuse. In the case of a short circuit of the coupler (capacitive divider), the entire switchgear would be simply damaged.

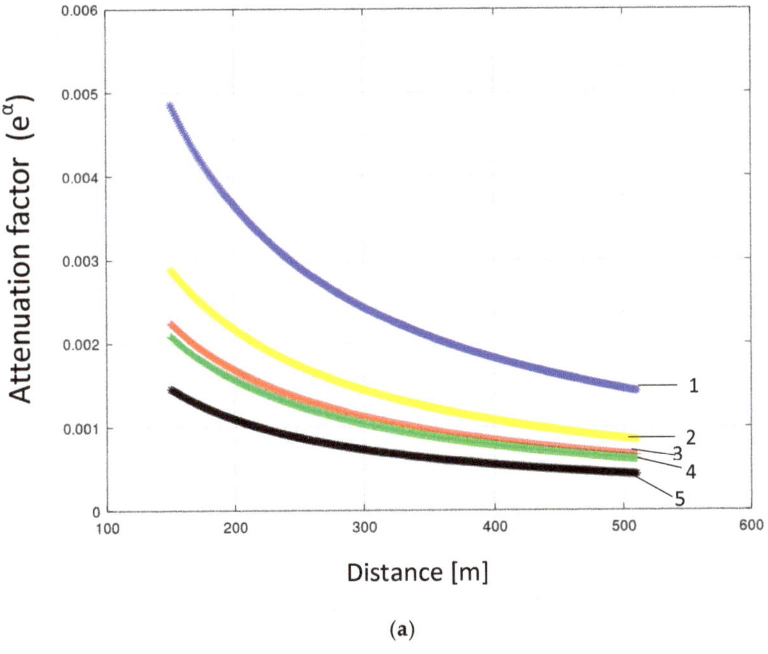


(a)

**Figure 5.** *Cont.*

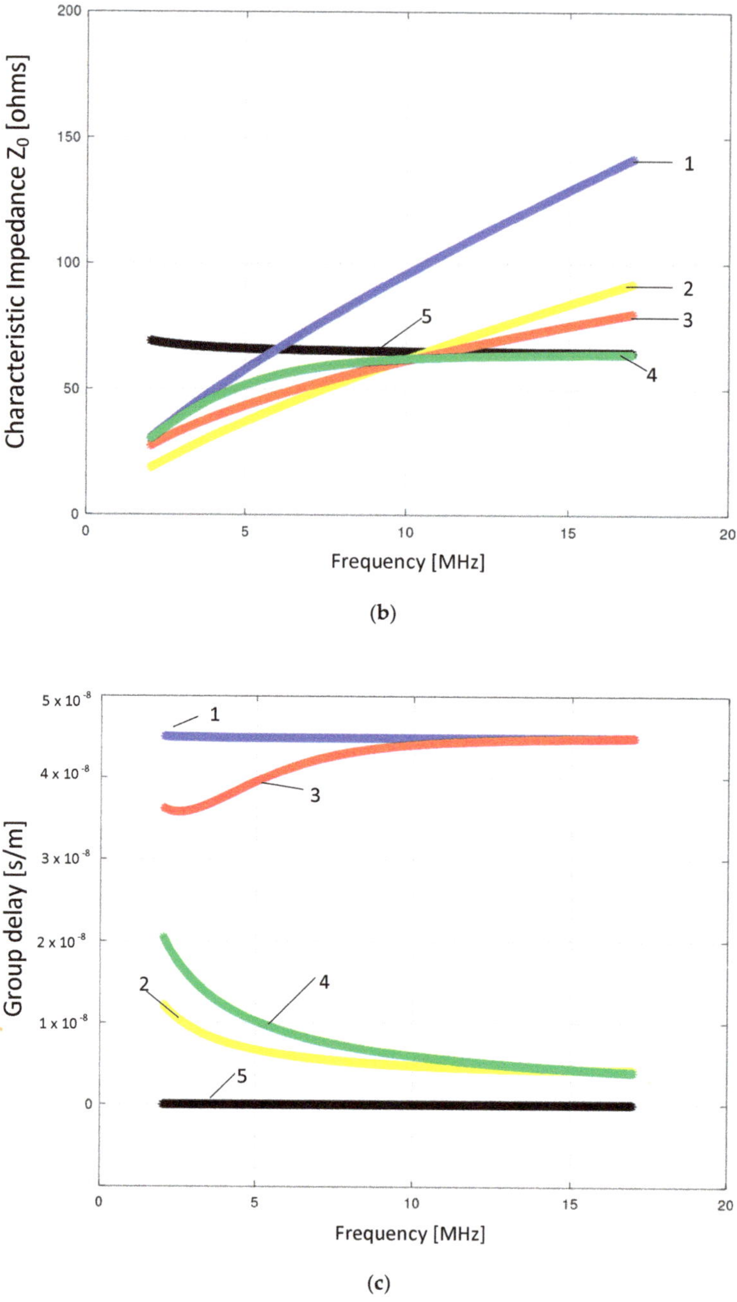


**Figure 5.** The impact of the cable's electrical parameters and frequency for the capacitive-inductive coupling: (**a**) Comparison of calculated values of attenuation factor $\alpha$ as a function of cable length in 5 analyzed approaches; (**b**) Calculated characteristic impedance $Z_0$ of the analyzed cable as a function of frequency in 5 analyzed approaches; (**c**) Calculated group delay versus frequency in 5 analyzed approaches.

According to obtained results, it seems that method 5 (approach no. 5) is also the most convenient for analysis of the capacitive-inductive coupling, from a practical point of view. From the attenuation waveforms, as shown in Figure 5a, and characteristic impedance, shown in Figure 5b, a relationship can be deduced. In this case, one should emphasize that the attenuation increases as a function of cable length. However, the value of characteristic impedance does not increase and stabilizes at around 60 Ω. When it comes to group delay, as shown in Figure 5c, method 5 (approach no. 5) also proved its superiority, compared to especially approach no. 1 and no. 3. However, these facts must be confirmed in detail by further examination, including a data packet length analysis.

### 4.2. Data Packet Length

The measurements for the TCP/IP protocol, depending on the length of the transmitted data packet, during a 10 min time interval, are shown in Figures 6 and 7.

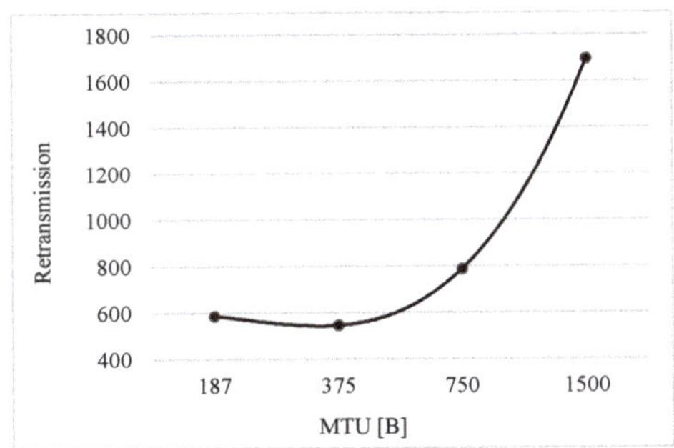


**Figure 6.** The number of retransmissions with respect to packet length.

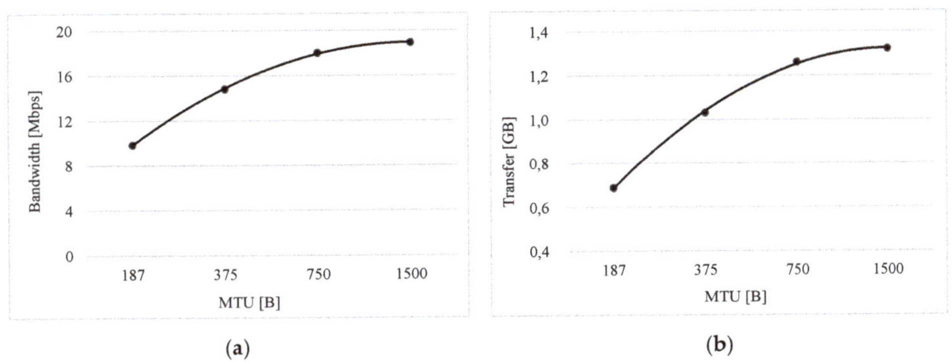


**Figure 7.** Dependency of packet length: (**a**) Network bandwidth; (**b**) Network transfer.

Based on carried out analyzes, it is clear that the smallest number of retransmissions occurs for small packet lengths. However, the use of small packets is closely related to the reduction of bandwidth at the TCP layer (measurements were taken without TCP window control). Obtained measurements show that reducing the length of a data packet significantly reduces the number of retransmissions in relation to the total number of transmissions, e.g., for 187 B, the number of retransmissions is equal to approx. 7%. However, for a packet of 1500 B, it increases to approx. 64%.

Subsequent measurements were therefore aimed at investigating the influence of the TCP window length on the packet retransmission effect. The results are shown in Figure 8. According to the analysis, it can be unequivocally stated that retransmission of packets does not occur for fixed TCP window lengths. However, the performance characteristics of the network depend on the length of the packet, adjusted to the size of the TCP window.

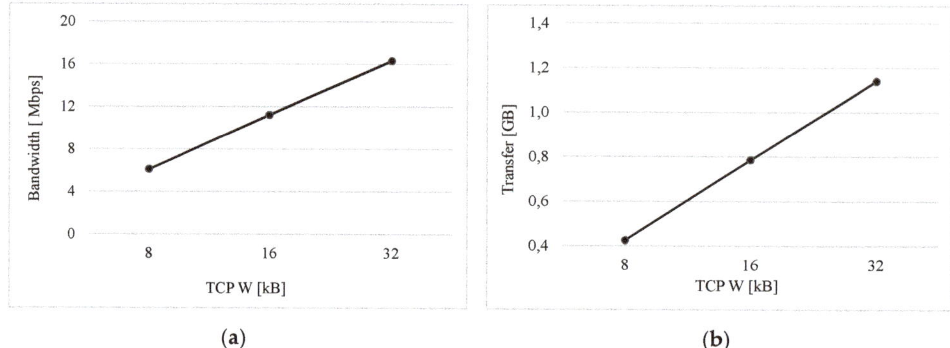


**Figure 8.** Dependency of TCP window length: (**a**) Network bandwidth; (**b**) Network transfer.

Next, for the UDP protocol, network delay measurements were carried out, depending on the length of the packet. Obtained results of delay time in the PLC network, for the selected cable, are shown in Figure 9.

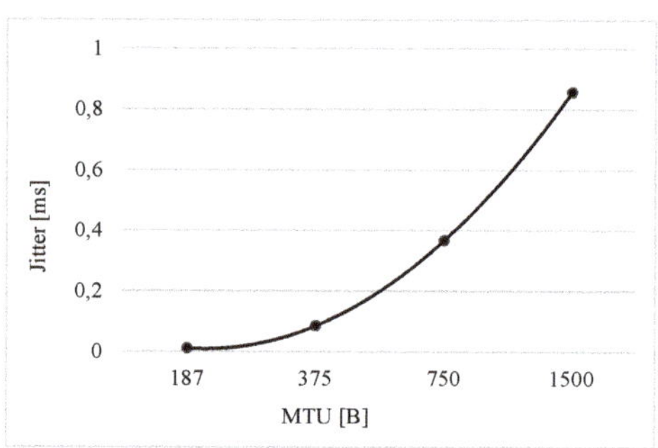


**Figure 9.** Dependency of network latency (jitter).

Based on obtained results, one can clearly see that the latency in the network reduces significantly with the decrease of packet length. Next, a subjective quality evaluation study has been carried out, to validate the efficiency and usefulness of a BPL-PLC voice communication system.

### 4.3. Voice Transmission

The voice transmission efficiency was determined for a real-time operating BPL-PLC system. Its main aim was to investigate how do varying conditions, concerning different types of coupling (inductive-inductive and capacitive-inductive), transmission mode (mode 1—frequency range of 3–7.5 MHz and mode 11—frequency range of 2–7 MHz), as well as bitrate (8, 16, and 24 kbps), affect the end-user quality.

The tested speech samples were sourced from ITU-T P.501 [28]. This dataset includes signal samples consisting of two sentences spoken by two female and two male individuals, in different languages. The length of each sample was equal to approx. 7 s. Taking into consideration the international character of the oil and mining industry, we have selected samples from 4 sets, namely: American English (AE), British English (EN), German (GE), and Polish (PL).

The original samples were available in the WAV 16-bit PCM format, with a sampling frequency set to 32 kHz. Whereas the degraded (transmitted) signal samples were coded at different bitrates, namely: 8, 16, and 24 kbps, using the Ogg Vorbis format, with sampling frequency changed to 44.1 kHz. We have decided to use this codec due to its openness and full compatibility with the Linux operating system, as our custom-designed communication network was running on Linux-powered devices. We intended to have as much control over the hardware and software layer as possible.

After being transmitted in real-time (from point A to B, and vice versa), all speech samples were recorded on each receiving side for further processing and evaluation purposes. Additional information on low bitrate audio coding may be found in [29–31]. Whereas the matter of consumption of multimedia content among end-users is discussed in [32].

The subjective user assessment was carried out using high-quality headphones in a 5-step mean opinion score (MOS) absolute category rating (ACR) scale, with no reference signal available, ranging from 1 (bad quality) to 5 (excellent quality). The study involved a group of 16 people, aged between 25–35 years old, which best match the profile of professionals from the oil and mining industry. Each participant assessed the quality individually, according to [33], and took a training phase before starting the essential study. A single session took approximately 25 min, with a short break in the middle of the test.

The results of the subjective quality evaluation, with respect to spoken language, bitrate, as well as transmission mode, are shown in Figure 10. Obtained grades have been averaged, considering: type of coupling (both inductive-inductive and capacitive-inductive), speaker (both female and male individuals), and transmission route (both A to B and B to A), respectively. It is worth mentioning, that previous studies had shown that BPL-PLC transmission itself is asymmetrical and its efficiency, resulting in available bitrate, is higher for mode 11 (up to 34 Mbps) compared to mode 1 (up to 20 Mbps).

When statistically processing obtained data of all recordings and tests made with different speakers (male and female), having chosen a 95% confidence interval, the dispersion (mean ratio) was less than 10%. For the sake of clarity, it was not marked in the diagrams. Furthermore, it should be noted that neither individual had hearing disorders, and was fluent in both English and German, whereas Polish was their mother tongue.

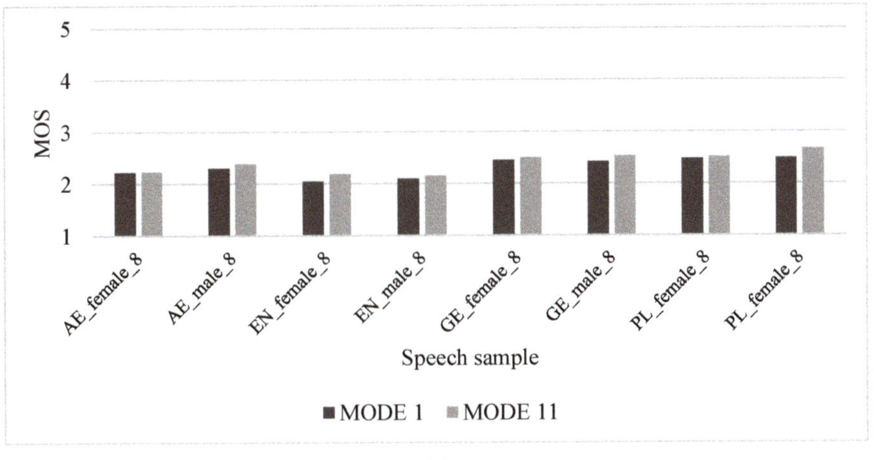


(a)

**Figure 10.** *Cont.*

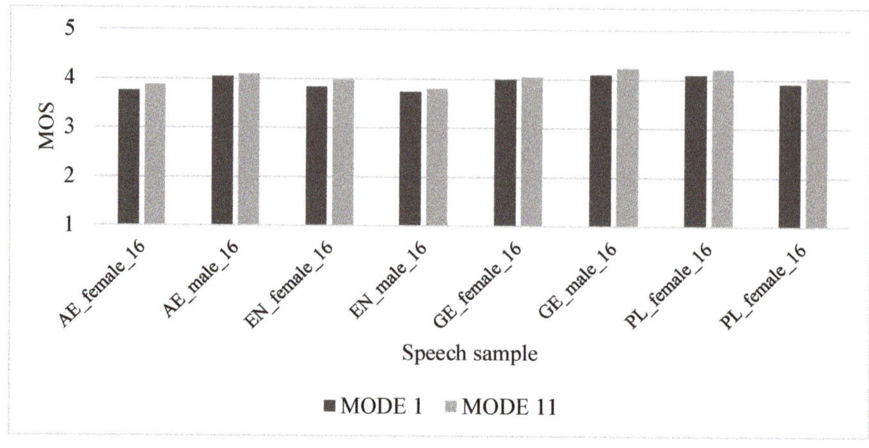


(**b**)

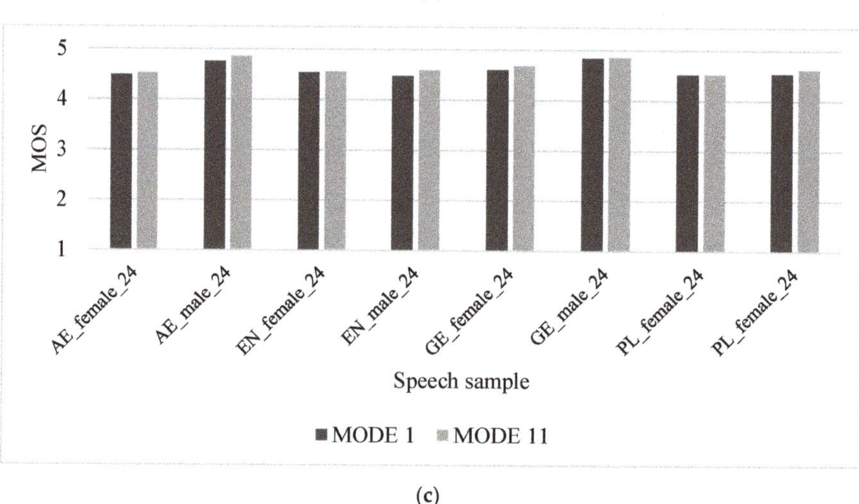


(**c**)

**Figure 10.** Averaged results of transmitted speech samples in American English (AE), British English (EN), German (GE), and Polish (PL), coded at different bitrates: (**a**) Bitrate of 8 kbps; (**b**) Bitrate of 16 kbps; (**c**) Bitrate of 24 kbps.

As shown, the lowest bitrate equal to 8 kbps proved to be insufficient when it comes to delivering easily understandable voice messages. The signal was distorted, noise, and crackling in the audio recording were clearly audible. On the other hand, the medium bitrate of 16 kbps was ranked evidently better. Nevertheless, despite the fact that all messages were understandable, they did not receive a good overall rating. In the case of the highest bitrate of 24 kbps, all voice messages, whether spoken by a male or female lector, were clear and easily understandable. Additional information concerning perceptual audio coding, psychoacoustics, as well as the impact of noise on hearing, may be found in [34–37].

When examining obtained results, one should take into account that any voice transmission system may be considered as high-quality whenever it receives a MOS score of above 4.0 (good quality). As shown, the threshold of 24 kbps may be viewed as a breakpoint, when a further increase in bitrate will not relate to further raise in perceived subjective quality. Additionally, this subjective study has also shown the superiority of transmission mode 11 over mode 1. In all cases, regardless of the bitrate or type of sample, speech signals were ranked evidently higher. This becomes an important factor, particularly when designing stable and reliable systems and services for the oil and mining industry.

## 5. Conclusions

The analysis, considering high-frequency signal propagation in the cable medium, had shown that the most useful method to provide quality parameters of BPL-PLC transmission, in hazardous areas with the use of the mining cable, is the method related with a variation of all electrical quantities of the cable with the signal frequency. The usefulness of coupling modems with the wired medium, both inductive-inductive and mixed capacitive-inductive, had been also confirmed. This fact is of great importance for applications, where there is no direct access to the phase, e.g., cables in a mine shaft. The positive impact of reducing the packet length, in the case of BPL-PLC transmission, was also demonstrated.

As shown, the BPL-PLC wired medium, thanks to its physical properties, can provide a reliable voice transmission system. Even in a narrowband scenario (bitrates lower than 1 Mbps), e.g., caused by bandwidth limitations and/or severe damage, etc., this technology ensures a stable and reliable connection. Whenever an emergency situation occurs, voice commands, i.e., from a supervisor or paramedic, can help during any rescue operation. Previous random events around the world have shown how important is to maintain contact and communication with cutout miners. Since voice can be transmitted effectively at 24 kbps, it may be assumed that a physical link with a speed of 1 Mbps possibly will be sufficient for this type of service. Therefore, such transmission parameters could theoretically correspond to a damaged cable. However, this particular case (a real damaged cable), has not yet been investigated by the authors. Whereas, an investigation focused on determining the impact of noise on the quality of voice transmission in a BPL-PLC communication system, may be found in [38].

It should be noted that the results of this work are innovative and not known previously in the literature. The technical medium examination had shown that inductive-inductive coupling has a clear advantage over mixed capacitive-inductive coupling. Moreover, when examining results in the case of voice transmission, there are situations in which mode 11 proved to be superior. As shown, the BPL-PLC technology can provide stable and reliable voice transmission services at 24 kbps, regardless of the spoken language, or even type of coupling. The results of this study can aid both researchers and scientists during the design and maintenance phase of a wired BPL-PLC voice communication system. Future studies may include research on the impact of aging on the physical parameters of the cable materials, as well as real mechanical damage to its elements, on the efficiency of voice transmission via the BPL-PLC communication system.

**Author Contributions:** Conceptualization, G.D. and P.F.-G.; methodology, G.D., P.F.-G., and M.H.; software, G.D. and P.J.; validation, G.W.; formal analysis, B.M.; investigation, G.D. and P.F.-G.; resources, J.W.; data curation, B.P.; writing—original draft preparation, G.D. and P.F.-G.; writing—review and editing, G.D., P.F.-G., B.M. and M.H.; visualization, G.D., P.F.-G. and M.H.; supervision, B.M.; project administration, A.W.; funding acquisition, A.W. All authors have read and agreed to the published version of the manuscript.

**Funding:** This research received no external funding. The APC was funded by General Tadeusz Kosciuszko Military University of Land Forces.

**Acknowledgments:** Authors would like to thank KGHM Polska Miedz S.A. and KOMAG Institute of Mining Technology for their support during this study.

**Conflicts of Interest:** The authors declare no conflict of interest.

## References


1. Meng, H.; Chen, S.; Guan, Y.L.; Law, C.L.; So, P.L.; Gunawan, E.; Lie, T.T. Modeling of transfer characteristics for the broadband power line communication channel. *IEEE Trans. Power Deliver.* **2004**, *19*, 1057–1064. [CrossRef]
2. Brown, P.A. Identifying some techno-economic criteria in PLC/BPL applications and commercialization. In Proceedings of the IEEE International Symposium of Powerline Communications and its Applications (ISPLC), Vancouver, BC, Canada, 6–8 April 2005; pp. 234–239. [CrossRef]
3. Carcelle, X. *Power Line Communications in Practice*, 1st ed.; Artech House: London, UK, 2009.
4. Uribe-Pérez, N.; Angulo, I.; De la Vega, D.; Arzuaga, T.; Fernández, I.; Arrinda, A. Smart grid applications for a practical implementation of IP over narrowband power line communications. *Energies* **2017**, *10*, 1782. [CrossRef]
5. Ikpehai, A.; Adebisi, B.; Rabie, K.M. Broadband PLC for clustered advanced metering infrastructure (AMI) architecture. *Energies* **2016**, *9*, 569. [CrossRef]
6. Kim, S.C.; Ray, P.; Reddy, S.S. Features of smart grid technologies: An overview. *ECTI Trans. Electr. Eng. Electron. Commun.* **2019**, *17*, 169–180. [CrossRef]
7. Pyda, D.; Habrych, M.; Rutecki, K.; Miedzinski, B. Analysis of narrow band PLC technology performance in low-voltage network. *Elektron. Elektrotech.* **2014**, *20*, 61–64. [CrossRef]
8. Mlynek, P.; Misurec, J.; Fujdiak, R.; Kolka, Z.; Pospichal, L. Heterogeneous networks for smart metering—Power line and radio communication. *Elektron. Elektrotech.* **2015**, *21*, 85–92. [CrossRef]
9. Debita, G.; Habrych, M.; Tomczyk, A.; Miedzinski, B.; Wandzio, J. Implementing BPL transmission in MV cable network effectively. *Elektron. Elektrotech.* **2019**, *25*, 59–65. [CrossRef]
10. Debita, G.; Falkowski-Gilski, P.; Habrych, M.; Miedzinski, B.; Wandzio, J.; Jedlikowski, P. Quality evaluation of voice transmission using BPL communication system in MV mine cable network. *Elektron. Elektrotech.* **2019**, *25*, 43–46. [CrossRef]
11. Debita, G.; Falkowski-Gilski, P.; Habrych, M.; Miedzinski, B.; Polnik, B.; Wandzio, J.; Jedlikowski, P. Subjective quality evaluation of underground BPL-PLC voice communication system. In *Theory and Applications of Dependable Computer Systems. DepCoS-RELCOMEX 2020. Advances in Intelligent Systems and Computing*; Zamojski, W., Mazurkiewicz, J., Sugier, J., Walkowiak, T., Kacprzyk, J., Eds.; Springer: Cham, Switzerland, 2020; Volume 1173, pp. 176–186.
12. Murphy, J.N.; Parkinson, H.E. Underground mine communications. *Proc. IEEE* **1978**, *66*, 26–50. [CrossRef]
13. Milioudis, A.; Andreou, G.; Labridis, D. Optimum transmitted power spectral distribution for broadband power line communication systems considering electromagnetic emissions. *Electr. Power Syst. Res.* **2016**, *140*, 958–964. [CrossRef]
14. Prasad, G.; Lampe, L.; Shekhar, S. In-band full duplex broadband power line communications. *IEEE Trans. Commun.* **2016**, *64*, 3915–3931. [CrossRef]
15. Fulatsa, Z.; Afullo, T.J.O. An alternative approach in power line communication channel modelling. *Prog. Electromagn. Res. C* **2014**, *47*, 85–93. [CrossRef]
16. Lampe, L.; Tonello, A.M.; Swart, T.G. *Power Line Communications: Principles, Standards and Applications from Multimedia to Smart Grid*, 2nd ed.; John Wiley & Sons: Hoboken, NJ, USA, 2016.
17. Lazaropoulos, A.G. Broadband transmission characteristics of overhead high-voltage power line communication channels. *Progr. Electromagn. Res. B* **2012**, *36*, 373–398. [CrossRef]
18. Dickinson, J.; Nicholson, P.J. Calculating the high frequency transmission line parameters of power cables. In Proceedings of the ISPCLA, Essen, Germany, April 1997; pp. 127–133.
19. Xiao, Y.; Zhang, J.; Pan, F.; Shen, Y. Power line communication simulation considering cyclostationary noise for metering systems. *J. Circuit Syst. Comp.* **2016**, *25*, 1650105. [CrossRef]
20. Artale, G.; Cataliotti, A.; Cosentino, V.; Di Cara, D.; Fiorelli, R.; Guaiana, S.; Panzavecchia, N.; Tinè, G.A. New coupling solution for G3-PLC employment in MV smart grids. *Energies* **2019**, *12*, 2474. [CrossRef]
21. Debita, G.; Habrych, M.; Wisniewski, G.; Miedzinski, B.; Tomczyk, A. Investigation and evaluation of coupling effect of BPL transmission units in MV cable network. In Proceedings of the International Symposium on Electrical Apparatus and Technologies (SIELA), Bourgas, Bulgaria, 3–6 June 2018; pp. 1–4. [CrossRef]
22. Carcangiu, S.; Fanni, A.; Montisci, A. Optimization of a power line communication system to manage electric vehicle charging stations in a smart grid. *Energies* **2019**, *12*, 1767. [CrossRef]

23. Barmada, S.; Tucci, M.; Fontana, N.; Dghais, W.; Raugi, M. Design and realization of a multiple access wireless power transfer system for optimal power line communication data transfer. *Energies* **2019**, *12*, 988. [CrossRef]
24. Kim, D.S.; Chung, B.J.; Chung, Y.M. Statistical learning for service quality estimation in broadband PLC AMI. *Energies* **2019**, *12*, 684. [CrossRef]
25. Hayt, W.H.; Buck, J.A. *Engineering Electromagnetics*, 6th ed.; McGraw-Hill: New York, NY, USA, 2001.
26. Sheen, D.M.; Ali, S.M.; Oates, D.E.; Withers, R.S.; Kong, J.A. Current distribution, resistance, and inductance for superconducting strip transmission lines. *IEEE Trans. Appl. Supercond.* **1991**, *1*, 108–115. [CrossRef]
27. Develi, I.; Kabalci, Y. Analysis of the use of different decoding schemes in LDPC coded OFDM systems over indoor PLC channels. *Elektron. Elektrotech.* **2014**, *20*, 76–79. [CrossRef]
28. ITU Recommendation P.501. *Test Signals for Telecommunication Systems*; ITU: Geneva, Switzerland, 2017.
29. Griffin, A.; Hirvonen, T.; Tzagkarakis, C.; Mouchtaris, A.; Tsakalides, P. Single-channel and multi-channel sinusoidal audio coding using compressed sensing. *IEEE Trans. Audio Speech Language Process.* **2011**, *19*, 1382–1395. [CrossRef]
30. Helmrich, C.R.; Markovic, G.; Edler, B. Improved low-delay MDCT-based coding of both stationary and transient audio signals. In Proceedings of the International Conference on Acoustics, Speech, and Signal Processing (ICASSP), Florence, Italy, 4–9 May 2014; pp. 6954–6958. [CrossRef]
31. Lin, J.; Min, W.; Shengyu, Y.; Ying, G.; Mangan, X. Adaptive bandwidth extension of low bitrate compressed audio based on spectral correlation. In Proceedings of the International Conference on Intelligent Computation Technology Automation (ICICTA), Nanchang, China, 14–15 June 2015; pp. 113–117. [CrossRef]
32. Falkowski-Gilski, P. On the consumption of multimedia content using mobile devices: A year to year user case study. *Arch. Acoust.* **2020**, *45*, 321–328. [CrossRef]
33. ITU Recommendation BS.1284. *General Methods for the Subjective Assessment of Sound Quality*; ITU: Geneva, Switzerland, 2003.
34. Kotus, J.; Czyzewski, A.; Kostek, B. Evaluation of excessive noise effects on hearing employing psychoacoustic dosimetry. *Noise Control Eng. J.* **2008**, *56*, 497–510. [CrossRef]
35. Herre, J.; Dick, S. Psychoacoustic models for perceptual audio coding—A tutorial review. *Appl. Sci.* **2019**, *9*, 2854. [CrossRef]
36. Esposito, A.; Amorese, T.; Cuciniello, M.; Riviello, M.T.; Esposito, A.M.; Troncone, A.; Cordasco, G. The dependability of voice on elders' acceptance of humanoid agents. In Proceedings of the Annual Conference of the International Speech Communication Association (INTERSPEECH), Graz, Austria, 15–19 September 2019; pp. 31–35. [CrossRef]
37. Meng, Z.; Gaur, Y.; Li, J.; Gong, Y. Speaker adaptation for attention-based end-to-end speech recognition. In Proceedings of the Annual Conference of the International Speech Communication Association (INTERSPEECH), Graz, Austria, 15–19 September 2019; pp. 241–245. [CrossRef]
38. Debita, G.; Falkowski-Gilski, P.; Habrych, M.; Miedzinski, B.; Polnik, B.; Wandzio, J.; Jedlikowski, P. Subjective and objective quality evaluation study of BPL-PLC wired medium. *Elektron. Elektrotech.* **2020**, *26*, 13–19. [CrossRef]

MDPI
St. Alban-Anlage 66
4052 Basel
Switzerland
Tel. +41 61 683 77 34
Fax +41 61 302 89 18
www.mdpi.com

*Energies* Editorial Office
E-mail: energies@mdpi.com
www.mdpi.com/journal/energies

www.ingramcontent.com/pod-product-compliance
Lightning Source LLC
LaVergne TN
LVHW070636170726
843515LV00004B/260

* 9 7 8 3 0 3 6 5 0 4 7 0 4 *